REPRODUCTIVE AGENCY, MEDICINE AND THE STATE

Fertility, Reproduction and Sexuality

GENERAL EDITORS:

David Parkin, *Director of the Institute of Social and Cultural Anthropology, University of Oxford.*

Soraya Tremayne, *Co-ordinating Director of the Fertility and Reproduction Studies Group and Research Associate at the Institute of Social and Cultural Anthropology, University of Oxford; and a Vice-President of the Royal Anthropological Institute.*

Volume 1
Managing Reproductive Life: Cross-Cultural Themes in Sexuality and Fertility
Edited by Soraya Tremayne

Volume 2
Modern Babylon? Prostituting Children in Thailand
Heather Montgomery

Volume 3
Reproductive Agency, Medicine and the State: Cultural Transformations in Childbearing
Edited by Maya Unnithan-Kumar

REPRODUCTIVE AGENCY, MEDICINE AND THE STATE

Cultural Transformations in Childbearing

Edited by
Maya Unnithan-Kumar

Berghahn Books
NEW YORK • OXFORD

First published in 2004 by
Berghahn Books
www.BerghahnBooks.com

©2004 Maya Unnithan-Kumar

All rights reserved. Except for the quotation of short passages for the purposes of criticism and review, no part of this book may be reproduced in any form or by any means, electronic or mechanical, including photocopying, recording, or any information storage and retrieval system now known or to be invented, without written permission of the publisher.

Library of Congress Cataloging-in-Publication Data
Reproductive agency, medicine and the state : cultural transformations in childbearing
  edited by Maya Unnithan-Kumar.
    p. cm.
  Includes bibliographical references and index.
  ISBN 1-57181-648-8 (alk. paper)
   1. Human reproductive technology – Social aspects – Cross-cultural studies.   2. Human reproductive technology – Government policy – Cross-cultural studies.   3. Human reproductive technology – Public opinion – Cross-cultural studies.   4. Childbirth – Social aspects – Cross-cultural studies.   5. Fertility, Human – Social aspects – Cross-cultural studies. I. Unnithan-Kumar, Maya, 1961–

RG133.5.R456 2004
306.4'61–dc22   2003063586

British Library Cataloguing in Publication Data
A catalogue record for this book is available from the British Library

ISBN 1-57181-648-8 hardback

Contents

List of Figures and Tables	vii
Preface and Acknowledgements	viii
Introduction: Reproductive Agency, Medicine and the State *Maya Unnithan-Kumar*	1
1. Attitudes to Genetic Diagnosis and to the use of Medical Technologies in Pregnancy: Some British Pakistani Perspectives *Alison Shaw*	25
2. Localising a Brave New World: New Reproductive Technologies and the Politics of Fertility in Contemporary Sri Lanka *Bob Simpson*	43
3. Conception Technologies, Local Healers and Negotiations around Childbearing in Rajasthan *Maya Unnithan-Kumar*	59
4. Programmes of Gamete Donation: Strategies in (Private) Clinics of Assisted Conception *Monica M. E. Bonaccorso*	83
5. Women, Doctors and Pain *William Stones*	103
6. Labour, Privatisation, and Class: Middle-Class Women's Experience of Changing Hospital Births in Calcutta *Henrike Donner*	113
7. In Search of Closure for Quinacrine: Science and Politics in Contexts of Uncertainty and Inequality *Asha George*	137
8. 'She Has a Tender Body': Postpartum Morbidity and Care during Bananthana in Rural South India *Asha Kilaru, Zoe Matthews, Jayashree Ramakrishna, Shanti Mahendra and Saraswathy Ganapathy*	161
9. 'And Never the Twain Shall Meet': Reproductive Health Policies in the Islamic Republic of Iran *Soraya Tremayne*	181
10. Women in Fertility Studies and *In Situ* *Tulsi Patel*	203

11. Heteronomous Women? Hidden Assumptions in the   223
    Demography of Women
    Sumi Madhok

Notes on Contributors   245

Index   249

List of Figures and Tables

Figures

Figure 8.1	Characteristics of the Study Population	166
Figure 8.2	Women Reporting Morbidities by Postpartum Duration	173
Figure 9.1	Population Pyramid Based on 1996 Census	187
Figure 9.2	Distribution of Unplanned Pregnancy by Parity in 1996	187

Tables

Table 8.1	Traditional Bananthana Practices by Duration of Postpartum Period	168
Table 8.2	Number of Providers Contacted for any Postnatal Care with and without Tubectomy-Related Contacts	170
Table 8.3	Results of a Logistic Regression to find Correlates of Routine Postnatal Contact	172
Table 8.4	Content of Routine Postpartum Care by Type of Health Care Provider	172

Preface and Acknowledgements

This volume is based on a workshop held at the University of Sussex in 2001 on the theme 'Anthropology, Reproduction and Health Policy'. The workshop was organised primarily as a means of bringing together some of us in the U.K. working in the area of human reproduction to exchange views and 'talk across' our disciplines. A number of common concerns emerged during the workshop: with agency, knowledge and the exercise of power, which also underlie this volume. Despite its interdisciplinary leanings, the volume is nevertheless shaped by an anthropological concern with analytic ethnography.

There are a number of people to thank: the contributors to this volume for sharing their work and good spirit, and colleagues and friends at Sussex who took time to attend and contribute to the workshop, Ann Whitehead, Hilary Standing, Ralph Grillo, James Fairhead, Ben Soares, Charles Abraham and Vinita Damodaran. A special thanks to Soraya Tremayne, Alison Shaw, Josephine Reynell Macdonald, Karina Kielmann and Jeannette Edwards for their careful reading and sound advice on parts of the volume. Barbara Bodenhorn, Caroline Humphrey, Marilyn Strathern, Stacy Pigg have all contributed important insights, though they may be unaware of this! For the funds for the workshop I am grateful to the Wellcome Trust. For the production of this volume I am indebted to Jan Brogden, Sue Emberton and Laura Griffith for help with editing, Paul Allpress and the computing unit for the timely loan of a laptop, and to Marion Berghahn for her supportive role in the publication of this volume.

Introduction: Reproductive Agency, Medicine and the State

Maya Unnithan-Kumar

The main objective of this volume is to consider the relationship between human reproductive processes (including attitudes to fertility, pregnancy, childbirth and the postpartum period), medical technologies and state health policies in diverse cultural contexts, especially outside Northern Europe and North America. Bringing together researchers from several disciplines, the volume discusses the relationship between local and global ideas, practices and policies concerning reproduction and health across the developing and post-industrial worlds. It seeks to understand the connections between biological and social reproduction: between how the physical processes of childbearing are connected to the reproduction of social institutions and values. The contributions are connected by a common interest in examining the exercise of medical power and the role of state policies and programmes to do with reproduction and health. The concept of 'reproductive health' provides a means of exploring various epistemological positions, and of understanding state ideas and practices relating to planned social change. At a more local level, a focus on the reproductive agency of women and men enables social and cultural responses to the processes of modernisation, in particular to the increasing intervention of biomedicine and reproductive technologies in people's daily lives, to be explored. The focus on reproductive technologies is pertinent in this context because 'reproductive technologies crystallise issues at the heart of gender, reproduction and family relationships and give insight into the engagement with modernity' (Stanworth 1987: 4). Examining the issues raised by reproductive technologies provides a useful insight into the ways in which people understand themselves to be connected (Strathern 1992a, Ginsburg and Rapp 1995, Edwards et al. 1993, Edwards 2000, Becker 2001, Ragone 2000).

Since the late 1980s, there has been a surge in scholarship on the relationship between human reproduction, biomedical technologies, and the related area of childbirth (Martin 1987, Stanworth 1987, Petchesky 1987,

Strathern 1992a, Edwards et al. 1993, Ginsburg and Rapp 1995, Davis-Floyd and Sargent 1997, Lock and Kaufert 1998, Davis-Floyd and Dumit 1998, Franklin and Ragone 1998, Rapp 2001, Edwards 2000, Inhorn 1994, 2000, Inhorn and Van Balen 2002). In anthropology, much of the scholarship on human reproduction was previously 'narrowly cast within the form of androcentrism, ethnocentrism and biological determinism that greatly limited the ways in which reproduction could be studied or analysed' (Franklin and Ragone 1998: 2). As a result, reproduction has only recently become central to social theory (Rapp, 2001: also see Weiner, 1995 for a similar concern).[1] Rapp notes that medical anthropology in particular has benefited from the feminist interrogation of problematic fertility and childbearing, as well as the more general critiques of the mind/body distinction. The current volume engages with these emerging perspectives, but it also addresses certain gaps in the literature.

Firstly, acknowledging that anthropology is not the only discipline to undertake empirically based and critically reflective methodological work in the area of human reproduction, the volume brings together demographic, public health, clinical and political science perspectives alongside anthropological ones to understand fully the local, academic and policy aspects relating to human reproduction. Approached through different disciplinary perspectives, a critical analysis of human reproduction, sensitive to hierarchical relations of power, serves, in turn, to bridge the gaps between, for example, anthropology, public health and demography (also see, for example, Greenhalgh 1995, Kertzer and Fricke 1997, Kielmann 2002, Ravindran and Panda 2002). Secondly, there are still relatively few studies on the engagement with reproductive technologies such as the ultrasound scan, assisted conception techniques, caesarian sections or genetic risk testing in non-Northern, Euro-American and especially 'developing/southern' countries, and amongst people from these regions (the recent work of Morsy 1995, 1998, Gruenbaum 1998, Boddy 1989, 1998, Kielmann 1998, Inhorn 2000, 2002, based mainly in Egypt and Sub Saharan Africa are notable exceptions). This volume aims to move beyond the dominant Northern Euro-American setting of the scholarship on medicalisation[2] to get a broader view of the cultural responses to the reproductive technologies and reproductive health programmes in Sri Lanka, India, Iran, Italy and amongst Pakistani groups in Britain. Going beyond the prevailing models of Euro-American kinship and citizenship, for example, we find that contestations around reproductive technologies take a different form and invoke different understandings of the ways people see themselves as connected.

In the following lines I first explain the three central themes: state power, medical control and reproductive agency, which inform this volume. I then discuss how these are addressed in the different chapters, and contribute to understanding the role of authoritative knowledge, emotions, kinship ideologies, development discourse, fertility, ethics and rights issues in reproductive change.

The State

State operations in relation to reproduction vary both in terms of the controls exercised upon its citizens and in terms of the reproduction of state values, subjects and institutions. By 'state' I mean 'both an illusory as well as a set of concrete institutions; as both distant and impersonal ideas as well as localised and personified institutions; as both violent and destructive as well as benevolent and productive' (Hansen and Stepputat 2001; 5). In 'southern' countries such as Sri Lanka, India and Iran, which are represented in this volume, development paradigms frame social realities for the people as well as for the state (see section on the state, childbearing and development, below). These, in turn, shape the state's actual and perceived approach to reproduction and, linked to this, people's experiences of governance (Foucault 1978, Moore 1999, Hansen and Stepputat 2001, Gupta 2001) as mediated through the arena of reproduction. Thus, when the issue of fertility is addressed at the level of the state and development organisations, it gains meaning beyond the concerns and anxieties of individuals and their families. The state is both galvanised and controlled by the global discourse on reproductive health and, at the same time, it uses it to produce and reproduce its own mechanisms of status and control. These processes, in turn, have important mental and physical implications for individual reproductive experiences and ideas of citizenship. Apart from the authority asserted by the medical establishment, poorer men and women, especially, must contend with a much more intrusive surveillance from state health officials, as well as with the power of development organisations.[3]

Where population planning and state medical programmes tend to converge, the national concern with the fertility of women in southern states results in experiences of medicalisation very different from those in Northern Euro-American contexts, where medicine and development are more distinct. As Foucault (1973, 1978) suggests, population classification exercises and medical advancement are both expressions of the power of the ideological controls that modern institutions have of the body; in combination, their control is overpowering. Embedded, as they are, within processes of reproduction, women's bodies become key targets of medical advancement, and embody and signify the power of science in late capitalist societies (Martin 1987). In developing countries, women also become key targets of population planning exercises, signifying the 'economic and social progress' of the nation (see overview section below).

Medicalisation

A central and contested issue in feminist scholarship on the relationship between women and technology has been the extent to which women are empowered by their resort to technological interventions. Popular belief, supported by medical interest, has tended to associate reproductive technologies with the attendant *choices* that they facilitate, therein enhancing women's control over their own reproductive processes. Yet increasing medical intervention in the area of birth has actually led women to experience a lack of control and a 'fragmentation' of their bodies, generated by the biomedical focus on the partial body or the body in terms of its parts (Martin 1987, Stanworth 1987). Reproductive technologies act as symbols, reflecting the ways in which birth and reproduction remain anchored in patriarchal and late capitalist ideologies, which continue to devalue women's role in social and biological reproduction. As Martin argues, however, such a situation does not preclude women's ability to 'manage their bodies in resistance to prevalent forms of thought that would impose certain regimens of time and behaviour' (according to Strathern 1992a: 67). In other words women's responses to medical intervention in biological reproduction can be quite complex. Feelings of empowerment are themselves connected with women's reproductive consciousness, which is a product of their 'social and biological constitution' (age, class, sexual preference) and individual circumstances (Petchesky 1987: 73).

In their volume on body politics, Lock and Kaufert (1998) suggest that one way of characterising women's complex responses to medicalisation is to describe them as 'pragmatic'. According to Lock, this means that 'individuals are not inevitably made into victims of medical ascendancy (although this clearly happens at times) but act most often on what is perceived by them to be in their own best interests' (2001: 481). Thus, Lock and Kaufert suggest that the dominant response of women to medicalisation is that of ambivalence coupled by pragmatism (1998: 2). The present volume supports this position, and at the same time reveals contexts in which medical authority and intervention are sought with deliberation and experienced as empowering. The chapters here highlight a spectrum of people's engagement with reproductive science and technology, from those who seek it unreservedly as in fertility treatment in Italy or caesarean section births among the middle class in Calcutta, to those who engage fearfully or ambiguously with such technologies, as in the responses to genetic counselling among British Pakistanis or in the case of poor women in rural Rajasthan; and finally to a public disengagement in their use, as found in the ethical debates surrounding Quinacrine sterilisations.

The fact that women may also feel empowered in their engagement with reproductive technologies leads us to focus on the role of physicians in these processes. Although clinicians and health workers play an important role in determining the nature of clients' engagement with medical technologies, their role in enforcing compliance to medical regimes cannot be straightforwardly assumed (see for example, Good 1993 et al., Good 1994). Physicians may also support clients' views and desires against the dictat of the state. Private gynaecological services in India, for example, present clients with opportunities to enhance their child bearing potential, thus helping them to resist state pressure to control their fertility, as may also be the case in Northern Euro-American contexts. In this context, it becomes important to consider the doctors' own perceptions of medicalisation and their agency in relation to the desires of their patients. Lock (2001), for example, documents the refusal of physicians to carry out tests for a recent gene associated with Alzheimer's disease, despite pressure from families that this be done. Furthermore, the differences in the culture of medical practice (such as the differences between the culture of public medicine and private medicine in India, or in the ways biomedicine is itself practised in the Indian context) may lead to different experiences of medical control.[4]

Aside from the role of biomedicine in governance and the routinisation of state discipline, recent feminist analysis has also critically reflected upon its significance in social reproduction more generally. Haraway (1993), Franklin and Ragone (1998), Rapp (1999, 2001), Ragone and Winddance Twine (2000), for example, show how current medical practices and ideas reproduce differences of gender, race, class, nationality. While the technological engagements described in this volume often have a class dimension to them, the effects of class are almost always qualified by caste, kin and gender ideologies in the contexts which these chapters describe. The material in this volume also challenges us to think about medicalisation in contexts where there are different conceptions of body and society, and where, for example, there may be a more collective as opposed to individual sense of 'ownership' of the body. This means exploring the different ways in which bodies and society are connected in different cultures. As Strathern (1992a: 76) observes in relation to the people of North Mekeo in central Papua New Guinea, bodies are not owned by persons because they are constituted by social relations. When reflected in body imagery, this suggests both social disjunction as well as conjunction, rather than the body as an integrated social whole. Petchesky (1995), on the other hand, shows how the Western idea of the body as individually owned is a recent development and a product of the dominance of Lockean thought which favoured individual over collective notions of property.

The term 'collective ownership', as I suggest in the volume, refers to those instances where the body is primarily constituted through others. Here the connection between notions of body and self involves the mediation of, and negotiation with, others in the social group. In such contexts, the introduction of the techniques of assisted conception may not result in separating women's bodies from their selves, or in conferring an individual identity on the foetus, as they would in societies where bodies are considered to be individually owned (Petchesky 1987, Strathern 1992a, 1992b, Franklin and Ragone 1998, Taylor 1998, Layne 1999). The reproductive technologies in Euro-American contexts have contributed to shaping perceptions of mothers/women and foetuses as distinct entities, even pitting the rights of one against the other (as reflected in the traumas of women who face abortion, for example). Such divisions do not necessarily arise outside these contexts partly because of the way gender dichotomies and reproductive processes in particular, are cross cut by notions of social obligation and responsibility engendered through the membership of social institutions such as class and caste. This leads us on to the question of the nature of women's reproductive agency in such relatively collectively shaped societies.[5]

Agency

Since Martin's (1987) study, there has been a continued emphasis on the need to think critically about the body, to continue to interrogate the assumption that the body is a fixed, material entity, subject to the rules of biological science. More recent thinking on the body suggests that it should be understood not as a constant amidst flux, but as the epitome of that flux (Csordas 1994); we should move beyond the tendency to treat the body as a passive entity upon which society imposes its codes towards an understanding of it as a source of agency and intentionality. This phenomenological approach to the body is concerned to develop a model of the body where the body is 'not only subject to external agency but also simultaneously an agent in its own world construction' (Lyon and Barbalet 1994: 48).

The concept of 'reproductive agency' is approached in this volume in terms of the ideas, actions, thinking and planning in the domain of human reproduction by women and men who engage in reproductive activities and seek healthcare services, as well as in terms of the strategies, compulsions and motivations which inform the actions of medical, clinical and health personnel. In the development literature especially, the notion of reproductive agency has tended to equate women in postindustrial countries with agency, while outside these contexts women are perceived as lacking in reproductive autonomy. However, a challenge to this dichotomy is implicit in the recent anthropological literature on the rising intervention of medicine in assisting procreation (Edwards et al. 1993, Franklin 1997, Becker 2001 and others). Contesting the romantic notion of women's reproductive freedom in the West, this challenge portrays, for example, the constraints and lack of choice and helplessness that in vitro fertilisation (IVF) seeking couples experience. Moreover, as the contributions to this volume suggest, the notion of agency itself has to be interrogated. Several chapters in this volume show a clear difference in strongly medically plural societies between seeking medical information, for example through ultrasound scans, and intending to *act* upon the information received. In other words, actions considered 'natural' following a particular technologically assisted medical diagnoses may often not take place, either because the goal of the exercise in the first place was simply to experience the technology and gain information, or because the diagnosis is used to confirm existing ideas, whatever its outcome. Thus, we can have 'reproductive agency' accompanied by a *lack* of visible action. The decision to forgo action, rather than action itself, also reflects autonomy, as Madhok argues in this volume. This assertion is further borne out in the contributions by Shaw, Unnithan-Kumar, Kilaru, Donner and Patel, whose work is about women who negotiate autonomy within conditions of structural subordination. This approach takes the very embeddedness of women's bodies as a focal point of understanding their agency.[6]

A Thematic Overview of the Chapters

Authoritative Knowledge and Reproductive Agency

Issues of power and agency are crucially connected to pre-eminent forms and practices of knowledge in society. In this volume, 'authoritative knowledge' (or the dominance of one knowledge system over another; Jordan 1997, Davis Floyd and Sargent 1997) is considered in the face of competing knowledges of reproduction and healthcare at the local level. The authority of medical personnel or of the scientific community is further complicated when, in medically plural societies, it is juxtaposed with other forms of knowledge and authority relating to reproduction and healing. The main questions which emerge from several of the chapters have to do with the framing of authoritative knowledge on health where there are competing knowledges of reproduction, such as technologically-informed knowledge, experientially-based knowledge and knowledge framed by religious and gender ideologies.

In thinking about the interweaving of biomedical and indigenous knowledge on childbirth the idea which emerges in this volume, is that 'indigenous' knowledge systems are not as closed or as bounded as they are often perceived to be. Shaw's chapter on Pakistani responses to genetic counselling in Britain, shows that reproductive decisions are the product of a complex interrelationship between locally based knowledge (as provided by spiritual healers and the experience of childbirth and healthcare in Pakistan and Britain), biomedical knowledge (represented by the genetic counsellors), and social and economic circumstances. The respondents in Shaw's study – the religious experts and the counsellors – all have differing ideas on the Islamic prohibitions relating to genetic testing and the termination of pregnancy. As Shaw's analysis suggests, even the religious interpretations may be modified by religious experts in response to an individual's circumstances.

Kilaru, Mathew, Ramakrishna, Mahendra and Ganapathy's contribution reveals overlaps as well as mismatches in biomedical and local views on postpartum healthcare, with biomedical prescription being followed only where it 'fits' in with indigenous knowledge (for example in the idea of 'rest' following childbirth). Kilaru et al. suggest that women's approaches to health seeking in the postpartum period are influenced by local knowledge about the accumulation and loss of water and blood in pregnancy and childbirth, by ideal type notions of the female body, and by the hot/cold humor based etiology of Ayurveda. In practice, this translates into constraints being placed on postnatal women's consumption of food and water, which contradict biomedical ideas about the importance of drinking fluids to facilitate breastfeeding. Unnithan-Kumar's chapter demonstrates the selective responses of local healers to biomedical techniques to facilitate conception. Local midwives respond positively to ultrasound scanning techniques, the results of which are used to confirm their own predictive power regarding the state of the foetus and, in turn, serve to raise midwives' prestige vis-a-vis other healers in the community. In so doing, the midwives appropriate women's reproductive experiences much in the same way as, Petchesky (1987) observes, men appropriate women's reproductive experiences in their use of visually linked technologies such as the ultrasound scan, enabling male doctors to reproduce not just babies but motherhood itself.

A key issue for several contributors to this volume is to understand how medical authority is constructed and asserted through the technologies of reproduction. Simpson describes the way that novel reproductive and genetic technologies in Sri Lanka are stimulating debates and interweaving local perspectives on health, body and personhood with contemporary western bioethical discourses. Clinicians in Colombo are caught between the opposing pull of a medical ethics which favours distantly related egg and sperm donors, and the logic of kinship traditions which prefer more closely related donors. In particular there appears to be a desire to use the husband's brother's sperm to achieve pregnancy. For some specialists, these local preferences were to be discouraged, while others believed, as did their clients, that allowing sperm to pass between known persons was the culturally appropriate way to approach the new technologies.

Both Bonacorso and Donner show how the doctors' interests and those of their clients are closely allied and serve to reinforce each other. In the cases of IVF treatment in Italy and caesarean section hospital births in Calcutta, physicians reflect the intensity with which women and couples seek out the relevant reproductive technologies. Bonacorso's chapter describes how clinicians working in private clinics for assisting conception simultaneously employ a highly medicalised language to impose their authority on their patients while using a 'language of common place' to empathise with their clients. The parallel use of two languages assists clinicians in their interaction with the couple and makes programmes of gamete donation, which are highly problematic because egg and sperm of third parties are used to achieve conception, a possible option. The two languages operate in complex ways: they shorten or create distance between clinicians and couples when necessary, and alleviate much of the tension and anxiety attached to the programme.

Donner's chapter describes how urban middle class women in Calcutta believe that amniocentesis, intra-uterine devices (IUD) and caesarean sections are medical technologies used by doctors to promote the health and welfare of mothers and children. These perceptions fit in well with the interests doctors have in the technical interventions which both enhance their authority and, at the same time, due to the great expense of these procedures, bring them good money. The local knowledge of healthcare provided by traditional midwives, on the other hand, is disappearing with the rise of hospital births where women are no longer 'allowed' to return to their natal homes after giving birth.

In his chapter, Stones, an obstetrician/gynaecologist working in the British health service provides a clinician's perspective on doctor–patient consultations. Focusing on the language of pain related to women's menstrual and chronic pelvic disorders, he discusses the boundaries between medical certainty and uncertainty. He suggests that sociocultural factors are critical in determining when both doctors and patients regard pain as clinically significant. George in her contribution to the volume, also interrogates the medical community's use of highly technical language. Taking the example of Quinacrine sterilisation, George focuses on the activities of specific scientists and doctors involved in research to promote it as a method of contraception. She suggests that Quinacrine researchers are able to influence their medical colleagues by presenting findings in scientific journals, technical fora and seminars, spaces that are rarely populated by critical voices from the non-medical community. George argues for a balance of power between the authoritative knowledge regimes of the scientists and its critics, which, if unchecked, always tend to favour the scientists. She emphasises that the scientific community should not be immune from analyses of how it achieves its authority, or of the social consequences of its pre-eminence.

Ambivalence and Desire in Human Reproduction

Differing responses to reproductive technologies are often the result of wider negotiations that invoke non-medical frameworks of procreation and are embedded within family and religious life as much as in previous experiences of healthcare. A significant point to emerge in a number of recent studies on reproductive technologies has been the ambivalence associated with the use of the technologies of assisted conception (Edwards et al. 1993, Franklin 1997, Lock and Kaufert 1998, Becker 2001, for example). The studies in this volume also suggest that there may be differing reactions to different reproductive technologies, with some engendering a more positive response than others. As Shaw describes in this volume, her respondents were well-disposed to the use of ultrasound technology, but unwilling to seek further technological intervention if the results showed the baby to be handicapped. We find that local notions of procreation were used to question and resist scientific authority and that rather than prepare for handling a disabled child, the respondents question the dominance and certainty of scientific knowledge.

Unnithan-Kumar describes a similar desire to seek scans as compared with any other technology among Rajasthani women. Further technological intervention is resisted due to a number of factors such as the fear of discriminatory treatment by doctors, fear of the pain accompanying biomedical treatment, notions of what it means to be a good wife and mother and, most significantly, a primacy accorded to spiritual agency in reproductive matters. As I have argued elsewhere (2001, 2002), women's choice of certain healers and certain techniques reflects a complex interrelationship between intimacy (the influence of those amongst one's kin whom one is intimate with) and notions of efficacy (healers whose cures are perceived to be effective) and depends on women's poverty and their social position (for example, the extent to which there are other women whose help can be called upon in the household).

The role of feelings and emotions[7] in shaping women's engagement with reproductive technologies and related health services emerges, for instance, in studies of assisted conception in the form of anxieties and fears of patients and relatives (Edwards et al. 1993, Franklin 1997, Layne 1999). However, there has been no direct focus on this in the Euro-American studies, and such a perspective is also absent from discussions of women's health in countries of the South. The work of cultural psychologists (Shweder and Levine 1984, Lutz and White 1986, Lutz and Lughod 1990, Shweder 1994, 2001, for example) and phenomenologists (such as Csordas 1994, Lyon and Barbalet 1994) could be valuably applied to an understanding of the ambivalences that surround women's recourse to biomedical intervention (Unnithan-Kumar 2003a).

Kinship Ideologies and Changing Practice

Recent scholarship on English and North American kinship[8] (Strathern 1992b, Layne 1999, Carsten 2000) shows how kinship ideologies and practices are embedded within wider social processes such as those of individualism, biological reductionism and consumerism. The recourse to reproductive technologies in these societies reflects this connection and is shaped by it. For example, in the idea that you can buy and sell babies or substances from which babies are created, or in the idea that you can engineer birth to suit your individual requirements, or in the construction of maternal altruism which underlie surrogacy arrangements (Cannell 1990, Strathern 1992a, Edwards et al. 1993, Ragone 1999). The studies of Ragone, Edwards and others suggest that the reproductive technologies both change ideas of kinship (such as when parenthood becomes fragmented into multiple social and biological aspects) and at the same time serve to maintain and reinforce the biological (genetic) basis of relatedness in Euro-American societies. For example, in her ethnographic work on surrogacy in the U.S., Ragoné (1994, 1999, 2000) suggests that the increasing popularity of gestational surrogacy (where use is made of donated ova) over traditional surrogacy (where the surrogate mother provides the ova) lies in the priority given by the commissioning parents for there to be no genetic connection between the surrogate mother and the child she gives birth to. So while surrogacy and donated ova are an increasingly accepted method of having children, the importance given to biological relatedness (in that the ova cannot be of the surrogate) continues to define who is regarded as a parent. The demand for such arrangements, as Ragoné describes, comes not only from the intending parents but also from the would be surrogate mothers.[9]

Simpson's chapter in this volume shows how the positive reaction to the reproductive techniques of assisted conception in Sri Lanka stem not only from a desire for 'western goods' but also because they are seen as a means of combating the violence following the recent political conflict within Sri Lankan society. New reproductive techniques are also seen as offering some redress to growing concerns, real or imagined, about the decline in male fertility due to changes in lifestyle, on the one hand, and high rates of abortion, on the other. He goes on to highlight the ways in which for the Sinhalese, aspects of the new technologies are linked to traditional kinship institutions of adoption and polyandry. That is, passing gametes and embryos between persons was regarded as similar to adoption, and receiving the husband's brother's sperm was like the traditional polyandrous relationship, where several brothers shared a wife. These ideas are very similar to Kahn's (2002) recent observations on kinship, 'adultery' and 'appropriate conception' as framing the ultra-orthodox Israeli Jews' acceptance of the IVF technologies. As Kahn suggests, the use of non-Jewish donor sperm to overcome the problem of 'adultery'[10] is a solution perfectly compatible with the ultra-orthodox Jewish notions of appropriate conception, as Jewishness is conferred through the matriline. The IVF techniques are also embraced by the Israeli state, according to Kahn, for they are percieved as a means of replacing community members lost to war at the same time as enabling Israel to maintain a parity with Arab and Palestinian birth rates.

The significance of local ideas of kinship in framing responses to the reproductive techniques is further evident in Bonacorso's chapter on IVF treatment in Italy. Medical doctors use the language and idioms of kinship to strengthen their client's belief in IVF techniques and outcomes. The language used evokes associations of 'naturalness' by connecting the present techniques to a past, non-medical, biological realm of child conception.

The desire for and use of reproductive techniques can be the source of disagreement and conflict within the family and between couples, as the bearing, birthing and nurturing of children is of central importance in gaining social recognition not only by the parental generation but also by grandparents and the wider kin group. Pakistani, Indian or other women in patrilineal contexts in South Asia, need not always be guided by their husband and his family's desires for children. At the same time, the husband may not always contradict his wife's desires not to have children. Women do act on their own initiative in matters of childbirth and this is further reflected in their uptake of reproductive technologies. However, there seems to be greater uncertainty attached to the outcomes when women contradict the wishes of their family than when they acquiesce to them as emerges in one of the cases in Shaw's chapter.

Intrakin conflicts can also be generated by a spatial relocation of the actual place where birth takes place. As Donner shows, the shift to birthing in hospitals associated with caesarian sections among the Calcutta middle-class radically alters the established obligations around birthing between the two sets of kin related by marriage. What is at stake is the right to provide and partake of postpartum care, and conflicts emerge in the tussle between the birthing woman's parents and her parents-in-law regarding where she should go from hospital. Increasingly, as Donner suggests, hospital births remove women's automatic rights to spend time in their natal homes after childbirth, thereby contravening established rules of kinship. Unnithan-Kumar suggests that the ways in which women respond to ultrasound techniques in Rajasthan is connected with their concern to demonstrate their fertility in response to the social value attached to childbearing and thus leads to a sense of reassurance and bonding with the affinal group rather than with the baby. At the same time, reproductive technologies help to interrogate patrilineal kinship ideologies, through for example, techniques which help counter what is widely regarded as women's errant or pathological reproductive conditions and behaviour, as seen in conditions of infertility.

The State, Childbearing and Development Discourse

Ideas on health and biological reproduction play a key role in processes by which the state (as represented by its various agents) is seen to reproduce its own institutions, values and attitudes. One important, recent concern within the anthropology of policy (for example, Shore and Wright 1997) is to consider not just how anthropological perspectives can contribute to the framing of policies but also the importance of anthropological reflection on the ways in which the policies of the government and non-government sectors alike construct their subjects (also see Grillo and Stirrat (1997) for a similar argument relating to an 'anthropology of development' that reflects critically on development discourse). Pigg (1997: 252, 253) demonstrates the practical consequences of health development discourse in its imposition of conformity on midwives to speak a specific language, at the same time as marginalising the importance given to local techniques and ideas of physiology. Focusing on midwifery training programmes in Nepal, she shows how the development practices of 'translation' (of local categories and knowledge) through which practitioners know, classify, manage and thereby 'develop' local traditions, serves to delink local practices of midwifery from their social context. This development practice results in widening the gap between planning processes and local realities, at the same time as presenting the development exercise as authoritative and beyond reproach.[11]

The centrality of midwifery programmes in the state agenda on health planning in itself reflects the significance placed on childbirth as defining women's reproductive experiences and needs. Such a perspective tends to de-emphasise other defining moments in the reproductive process more tangentially associated with childbirth, such as the onset of menstruation or its decline, the postpartum period which occurs just after birth, or conditions such as infertility which are associated with the inability to produce children. Kilaru, Mathews, Ramakrishna, Ganapathy and Mahendra argue in their chapter that in its focus on childbirth as central to reproduction, the Indian state is unable fully to acknowledge or address the postpartum risks faced by women who have given birth. The postpartum period has received little attention in maternal health planning and policy exercises, despite clear evidence that postpartum deaths in developing countries are more common than deaths during pregnancy and childbirth. In the postpartum phase, in contrast to the antenatal or delivery phase, the mother is physically separate from the child she has given birth to; her health is thus decoupled from the health of the child she has gestated, making it easier to ignore. The state's ignorance of the importance of the postpartum period contrast with the significance it is accorded by members of the southern Karnataka communities which Kilaru et al. studied, where the postpartum period (*bananthana*) is culturally well recognised and defined. Kilaru et al. call into question the mismatch between health policies which emphasise routine care to consist of tubectomies and immunisation, and the local realities of women's postpartum vulnerabilities. In some states in India, such as in Tamil Nadu, there is provision for postpartum care (Van Hollen 2002). Van Hollen describes the state's capacity to 'manoeuvre' and educate Tamil Nadu women in relation to their nutritional practices in the postpartum period, at the same time as constructing them as backward and even criminal (2002: 176). Thus, even in these contexts, we find that the state care provided is limited by specific notions of development that are at odds with local practices.

There is also an important, underlying class dimension to the practice of state health service delivery. Very often, as Pigg and Van Hollen's studies have pointed out (also see Ram 1998), it is poor women who are most subjected to the 'top-down' impositions of health officials. Poorer women are more likely to be subject to development intervention and medical coercion. Unnithan-Kumar (2003b) observes in Rajasthan these women are pressured by public health personnel to undergo tubectomies at the same time as having an abortion. Poor women may resist such impositions through their recourse to indigenous systems of care and this may lead them to be cautious of equating modernity with the use of recently available reproductive technologies. Women from more affluent families can choose the type of health intervention they prefer, and often use wider middle-class ideologies to negotiate along the lines of their desires. Donner's chapter in this volume describes, for example, how the equation of caesarean sections with 'progressiveness' is a key factor in its popularity among the Calcutta middle classes.

The processes by which state health policies and programmes marginalise, stigmatise or completely ignore locally important issues or even categories of people, is also the subject of Tremayne's chapter in this volume. Examining what are considered by many to be progressive population and family planning programmes of the Islamic Republic of Iran, Tremayne finds that the policies ignore the most sexually active and vulnerable group of Iran's population group, i.e., young people under twenty years, but more specifically under fifteen years. Going through the history and political economy of the region, Tremayne shows how state policies in a religious context can be simultaneously progressive and regressive. Political and economic realities constrain the religious leaders, as is reflected in the shifting nature of reproductive health policies. At the national level, religious beliefs are used to support the shifts in policy emphasis. So while, in the aftermath of the revolution, the Iranian clergy restrain birth control to encourage larger families, in the following decade they are seen as supporting birth control in order to promote economic prosperity, even arguing in terms of its roots in ancient Islamic practice.

In his chapter on the responses to the reproductive technologies in Sri Lanka, Simpson makes a similar argument about the connection between economic and political vulnerabilities of the state, and the politicians' and health planners' promotion of fertility. Simpson suggests that the growing sense of national vulnerability, as a result of the recent conflicts, is reflected in the anxieties around the reproduction of the family and around fertility, especially because of women's resort to abortion and the decline in male fertility. There is the fear that distortion of traditional arrangements for family reproduction will imperil the country's future. In Simpson's chapter we see clearly the parallel between the reproduction of culture, values and institutions of the state, and individual desires for biological reproduction.

Even communities living in post-industrial societies may have state agendas relating to reproduction imposed upon them, and may in turn have ways of negotiating or resisting such impositions. The reproductive practices and possibilities of, especially, the weaker, minority sections are prone to get caught up in the reproductive politics of the state. In the British Pakistani case, discussed in Shaw's chapter, an unintended result of genetic counselling discourse is that it can feed into the politics of race and immigration and further stigmatise the minority community of British Pakistanis deemed 'at risk' of genetic disorders through its marriage patterns. On the other hand, the connection between reproductive risk and migration is played out very differently in the British Pakistani discourse on racism, in which the perception is that there is greater 'risk' in the U.K. compared to Pakistan, where either the religious healers make people better, or fate ensures healthier families. Although the geneticists point to the fact that genetic disorders are likely to be present but less likely to be identified in Pakistan, the British Pakistanis whom Shaw interviewed do not necessarily share this perception.

Fertility, Women's 'Autonomy' and Reproductive Change

One of the ways in which the study of fertility aims to be connected to social realities – bringing the concerns of public health, anthropology and demography closer together – is through the link that is made between the potential to bear children and women's ability and freedom to make reproductive decisions. Further correlations are made in demographic, public health and development literature between women's ability to control their fertility and their overall health. However, it is less clear, in this literature, whether the primary aim is to control women's fertility, or to improve their health: the two issues may be less closely connected than is widely imagined. In population planning, the reduction in fertility and number of live births is a primary concern, but its connection with women's health may be more tenuous than health experts believe. In a recent study based on empirical findings from Tamil Nadu, southern India, for example, Ravindran and Panda (2002) question the premise of population transition theory that women's fertility and their overall social and health status are directly related. They suggest that increasing women's autonomy is not a necessary prerequisite for increasing women's use of contraceptives, and that fertility decline (as a result of the increase in use of contraception) does not necessarily lead to improvements in women's health.

Despite the global agreement on the significance of paying attention to women's health as defined by their own concerns, over and above the fertility driven planning processes of national governments, fertility and mortality rates continue to govern the health policies in developing countries such as India (this is not to deride the state and NGO-generated community initiatives in health). Health statistics, whether in the form of fertility and mortality rates or as part of Human Development indicators, continue to constitute authoritative knowledge in health policy and planning decisions (Kielmann 2002: 160) and fertility remains 'situated' within a typical demographic discourse (Greenhalgh 1995).

The issue of the demographic definition of 'autonomy' as related to women is at the heart of the chapters by Madhok and Patel in this volume. Both argue that the demographic understanding of autonomy has failed to pay attention to women's agency primarily because it equates autonomy with individualism. The subordinated conditions of women's lives in the developing world should not be taken as evidence of their lack of agency. In order to recover their agency, Madhok suggests we have to modify individualistic and 'act-centred' conceptions of autonomy. Accordingly, she argues that demographic and public health studies need to recognise autonomy capacities, and that in oppressive circumstances the autonomy capacity of individuals may not translate into actions. Drawing on her study of the village workers of the Women's Development Programme in Rajasthan, Madhok shows that in most cases Sathin women's desires and preferences on issues ranging from childbirth to political participation, are in contrast to their actions. The gap between the development of personal agency and ability to exercise agency needs to be acknowledged, through improved qualitative indicators and narrative methods. Patel, on other hand, argues for a more fluid understanding of social status in demographic literature. These studies complement Kielmann's (2002: 159) interrogation of the social relevance of public health assumptions, where she suggests that the variables used to measure women's status have rarely included discussions of how women themselves define health, how they perceive change and its impact on their well being, and how they situate themselves and respond to the multiple discourses on women's health to which they are exposed. She argues that perceived morbidity is an important indicator which can tell us far more about social change and its impact on health than do statistical representations.[12]

We question a further assumption in the literature on women's fertility in this volume which is the idea that women in developing countries are only concerned with controlling conception and have no interest in enhancing their childbearing potential (also see Inhorn and Van Balen 2002). The anti-natalist policies of governments in countries where human fertility is connected with the aims of population reduction often contrasts with the reproductive desires of the local people in these countries. In her chapter, George describes how the promoters of Quinacrine sterilisation believe that their method of sterilisation empowers women and addresses maternal mortality, in a process cheaper and easier than surgical sterilisation. However, in contrast to the Quinacrine promoters, access to contraception alone is not a direct solution to maternal mortality as pregnancies may be wanted and planned.

Medical Ethics and Reproductive Rights

The concept of 'reproductive health' as it is used in population-development programmes is a further example of the irony of development: where rights in the domain of conception and access to healthcare are acknowledged by the state and 'given' to its citizens, but are seldom 'taken' (or made use of) because of the social and economic inequalities which frame such access. The emergence of the concept of reproductive health is closely allied with the contribution of feminist critiques of development policies. Longstanding feminist and health activist concerns regarding the gender injustice of the population and family planning policies of governments were acknowledged only as late as 1994, at the International Conference on Population and Development. Feminists from the North and South argued that the state's control of female fertility could not be regarded as solely a demographic or economic issue but had to be connected to the issues of human rights and welfare (Correa 1994, Sen et al. 1994, Hartmann 1995, Petchesky and Judd 1998).

The issue of ethics around medical practice and state health interventions arises in the context of the concern with institutional accountability and the responsibilities and capabilities of those empowered by the healthcare programmes such as medical doctors and health officials and workers at various levels. In this volume, George's chapter highlights the complex issues and agendas of various institutional agents involved in the clinical trials to establish the efficacy and safety of Quinacrine as a female contraceptive. George stresses the need for large collective institutional mechanisms of accountability, especially significant given the rise of global pharmaceutical and medical networks against whom state controls may be ineffective. The global nature of inequalities generated by multinational corporations together with the cross-national authority of scientific knowledge significantly weakens the authority of the state as a player in the international politics of health.

In countries such as Iran and India (Tremayne and Kilaru et al. this volume) the state may be conscious of the need for regulation but is either caught up in promoting its own agenda or unable to regulate private medical interventions in health. As Rao (1999) and others argue for India, for example, the ethics of the state is compromised by foreign funders such as the World Bank. In the case of Italy, Bonacorso explains that there is a lack of legislation on assisted conception, except for an administrative act which establishes a general rule. The public sector (NHS) can provide any treatment in the field of assisted conception, except when third party gametes are used. The provision of treatment with third party gametes, which is highly controversial, is left entirely in the hands of the private sector.

These realities raise serious questions about safeguarding individual and community mechanisms of reproductive and bodily control. The role of feminist and other health activists becomes crucial in redressing miscarriages of justice in relation to reproduction and health, given the unequal context in which poor women, especially, receive medical care. In the long run, the issue of how reproductive rights are locally conceptualised, and the extent to which these ideas play a role in framing global discourse and national agendas on childbearing, health and development, will be critical (Petchesky and Judd 1998, Cornwall and Wellbourn 2002, Unnithan-Kumar 2003a). Recent anthropological studies make connections between knowledge and ethics, and explore conceptualisations of kinship in ways that inform medical ethics, policy and legislation surrounding the use of the health technologies (Strathern 1992a, Edwards et al. 1993). In the area of ethics and autonomy, Petchesky and Judd (1998), for example, have worked on reproductive rights from a feminist, activist and political theory framework.

Issues of power, knowledge and agency in relation to reproduction are emerging concerns in disciplines and sub-disciplines that include the anthropology of medicine and policy, ethnodemography, health studies, feminist studies, and legal studies.[13] The contributions to this volume complement and, in important ways, extend these emerging perspectives.

Notes

1. Weiner (1995) holds the lack of attention to reproduction as responsible for the theoretical differences between the more biologically orientated and the more culturally inclined feminists. Weiner argues that women's power should be seen as a synthesis of biological and cultural activities. Weiner suggests that a focus on reproduction enables a better understanding of kinship in particular and social theory in general. She emphasises the role of the political relations of biological reproduction in shaping western political and social histories.
2. Following Lock (1998, 2001), I use the word medicalisation to refer not only to the appropriation of women's bodies as a site for medical practice, but

also to the social arrangements and political forces that contribute to such experiences.
3. Edwards (1995) addresses a similar issue of the surveillance of poor men and women by the health services in her work in Northern England.
4. Also see Adams (2002) on how Tibetan doctors characterise 'outside medicine' as representing sinicised biomedicine in relation to their own practice of biomedicine.
5. I use the word relatively in describing the collective nature of many societies operating outside the frame of Anglo-Saxon ideologies because as Dumont (1966 [1980]) and Parry 1974 have shown, that despite the dominance of a collective orientation of the self in Indian society, individualism is nevertheless practiced and upheld as a value in certain contexts (also see Unnithan-Kumar 1997).
6. There is no doubt a focus on women's agency in this volume, nevertheless, men emerge as significant reproductive agents in their role as influential kinspersons (as husbands, brothers, fathers), as well as in relation to their position as custodians of authoritative knowledge (as doctors, healers, health officials).
7. In my use of the word emotion, I follow Richard Shweder's definition of 'emotion' as a complex notion wherein particular emotions of sadness guilt, envy are derived from various combinations of wants, beliefs, feelings and values (Shweder 2001).
8. I am aware that the term North American kinship is a very generalised one. In fact, recent work on kinship (Carsten 2000, Bodenhorn 2000) underscores the need to qualify such general notions. Bodenhorn's work on the Inupiat of Northern Alaska shows, for example, that for these Americans, biological ties, though valued, may be seen as optional rather than given and need to be considered in terms of the high value given to individual autonomy. Bodenhorn's work presents us with a contrary view to the singular 'essence' of American kinship as portrayed in Schneider's work.
9. The driving factors behind the demand for gestational surrogacy arrangements come from both the women who act as surrogates as well as from the intending parents. As Ragone (2000) observes, the surrogates prefer this arrangement because it is easier to undertake surrogacy when the child is 'not theirs'. this was especially the case in overcoming the 'problem' of race presented by interracial surrogacy arrangements. Parents also preferrred gestational surrogacy as it allowed them a greater choice of ova, and also because they believed in cases of conflict, that the courts were likely to grant custody of the child to the traditional as opposed to the gestational surrogate mother.
10. In order to overcome the problem of adultery caused by the artificial insemination of sperm of another man, a Jewish woman married to an infertile Jewish man is inseminated by non-Jewish donor sperm (Kahn 2002: 290)
11. Pigg powerfully argues that what are marginalised as merely custom or belief are in fact 'actual practices through which people care for bodies that are understood in terms other than those of biomedicine' (1997: 247). Her work emphasises the extent to which the idea of development as 'progress', as the idea of leaving behind traditional ways, is entrenched among the people and the state alike and reflected in practices surrounding conception and birth.

12. Kielmann supports Caldwell's position that the positive aspects of this health transition result from the interplay of 'modernising social forces' as well as direct 'biomedical interventions' (Kielmann 2002: 161)
13. In the legal field see, for example, the work of Sheldon (1997), also for an analysis of the recent legislation in the U.K. relating to men's reproductive concerns (1999).

References

Adams, V. 2002. 'Establishing Proof: Translating "Science" and the State in Tibetan Medicine', in M. Nichter and M. Lock, eds. *New Horizons in Medical Anthropology*. London: Routledge

Becker, G. 2001. *The Elusive Embryo: How Women and Men Approach New Reproductive Technologies*. Berkeley: University of California Press

Boddy, J. 1989. *Wombs and Alien Spirits: Women, Men and the Zar Cult in Northern Sudan*. Madison: University of Wisconsin Press

—— 1998. 'Remembering Amal: on Birth and the British in Northern Sudan', in M. Lock and P. Kaufert, eds. *Pragmatic Women and Body Politics*. Cambridge: Cambridge University Press, 28–58

Bodenhorn, B. 2000. '"He Used to be My Relative": Exploring the Bases of Relatedness among Inupiat of Northern Alaska', in J. Carsten, ed. *Cultures of Relatedness*. Cambridge: Cambridge University Press, 128–49

Browner, C. and Press, N. 1997. 'The Production of Authoritative Knowledge in American Prenatal Care', in R. Davis-Floyd and C. Sargent, eds. *Childbirth and Authoritative Knowledge*. Berkeley: University of California Press, 113–32

Cannell, F. 1990. 'Concepts of Parenthood: The Warnock Report, the Gillick Debate and Modern Myths'. *American Ethnologist*, 17: 4

Carsten, J. 2000. 'Introduction'. *Cultures of Relatedness: New Approaches to the Study of Kinship*. Cambridge: Cambridge University Press, 1–37

Cornwall, A. and Wellbourn, A., eds. 2002. *Realising Rights: Transforming Approaches to Sexual and Reproductive Wellbeing*. London: Zed Books

Correa, S. 1994. *Population and Reproductive Rights: Feminist Perspectives from the South*. London: Zed Books

Csordas, T. 1994. *Embodiment and Experience: The Existential Ground of Culture and Self*. Cambridge: Cambridge University Press

—— 2000. 'The Body's Career in Anthropology', in H. Moore, ed. *Anthropological Theory Today*. Cambridge: Polity, 172–206

Davis-Floyd, R. and Dumit, J. 1998. *Cyborg Babies: From Techno-Sex to Techno-Tots*. New York: Routledge

—— and Sargent, C. 1997. *Childbirth and Authoritative Knowledge: Cross-Cultural Perspectives*. Berkeley: University of California Press

Dumont, L. 1980. *Homohierarchichus: The Caste System and its Implications*. Chicago: University of Chicago Press

Edwards, J. 1995. 'Parenting Skills: Views of Community Health and Social Service Providers'. *International Journal of Social Policy*, 24: 2, 237–59

—— 2000. *Born and Bred: Idioms of Kinship and New Reproductive Technologies in England*. Oxford: Oxford University Press

—— Franklin, S., Hirsch, E., Price, F. and Strathern, M. 1993. *Technologies of Procreation: Kinship in the Age of Assisted Conception*. London: Routledge

Escobar, A. 1994. *Encountering Development: The Making and Unmaking of the Third World*. Princeton: Princeton University Press

Franklin, S. 1997. *Embodied Progress: A Cultural Account of Assisted Conception*. New York: Routledge

—— and Ragoné, H., eds. 1998. *Reproducing Reproduction: Kinship, Power and Technological Innovation*. Philadelphia: University of Pennsylvania Press

Foucault, M. 1973. *The Birth of the Clinic*. London: Tavistock

—— 1978. *The History of Sexuality*, Vol 1. London: Peregrine

Ginsburg, F. and Rapp, R. 1995. *Conceiving the New World Order: The Global Politics of Reproduction*. Berkeley: University of California Press

Good, B. and Good, M-J. Del Velcchio. 1993. 'Learning Medicine: The Constructing of Medical Knowledge at the Harvard Medical School', in S. Lindenbaum and M. Lock, eds. *Knowledge, Power and Practice*. Berkeley: University of California Press, 81–108

—— 1994. *Medicine, Rationality and Experience*. Cambridge: Cambridge University Press

Greenhalgh, S. 1995. *Situating Fertility: Anthropology and Demographic Enquiry*. Cambridge: Cambridge University Press

Grillo, R. and Stirrat, R., eds. 1997. *Development discourse: Anthropological perspectives*. Oxford: Berg

Gruenbaum, E. 1998. 'Resistance and Embrace: Sudanese Rural Women and Systems of Power', in M. Lock and P. Kaufert, eds. *Pramatic Women and Body Politics*. Cambridge: Cambridge University Press, 58–77

Gupta, A. 2001. 'Governing Population: The Integrated Child Development Services in India', in T. Blom Hansen and F. Stepputat, eds. *States of Imagination*. Durham: Duke University Press, 65–97

Hansen, T-B. and Stepputat, F. eds. 2001. 'Introduction'. *States of Imagination: Ethnographic Explorations of the Postcolonial State*. Durham: Duke University Press, 1–41

Haraway, D. 1993. 'The Biopolitics of Postmodern Bodies: Determinations of Self in Immune System Discourse', in S. Lindenbaum and M. Lock, eds. *Knowledge, Power and Practice: The Anthropology of Medicine in Everyday Life*. Berkeley: University of California Press, 364–411

Hartmann, B. 1995. *Reproductive Rights and Wrongs*. Boston: Southend Press

Hastrup, K. 2003. 'Representing the Common Good: The Limits of Legal Language', in R. Wilson and J. Mitchell, eds. *Rights, Claims and Entitlements*. Routledge ASA series (in press)

Inhorn, M. 1994. *Quest for Conception: Gender, Infertility and Egyptian Medical Traditions*. Philadelphia: University of Pennsylvania Press

—— 2000. 'Missing Motherhood: Infertility, Technology and Poverty in Egyptian Women's Lives', in H. Ragone and F. Winddance Twine, eds. *Ideologies and Technologies of Motherhood*. New York: Routledge

—— 2002, in M. Inhorn and F. Van Balen, eds. *Infertility around the Globe: New Thinking on Childlessness, Gender and Reproductive Technologies*. Berkeley: University of California Press

Jeffrey, R. and Jeffrey, P. 1997. *Population, Gender and Politics: Demographic Change in Rural India*. Cambridge: Cambridge University Press

Jordan, B. 1997. 'Authoritative Knowledge and Its Construction', in R. Davis-Floyd and C. Sargent, eds. *Childbirth and Authoritative Knowledge: Cross-Cultural Perspectives*. Berkeley: University of California Press

Kahn, S. 2002. 'Rabbis and Reproduction: The Uses of the New Reproductive Technologies among Ultra Orthodox Jews in Israel', in M. Inhorn and F. Van Balen, eds. *Infertility Across the Globe*. Berkeley: University of California Press, 283–98

Kertzer, D. and Fricke, T. 1997. *Anthropological Demography: Toward a New Synthesis*. Chicago: Chicago University Press

Kielmann, K. 1998. 'Barren Ground: Contesting Identities of Infertile Women in Pemba, Tanzania', in M. Lock and P. Kaufert, eds. *Pragmatic Women and Body Politics*. Cambridge: Cambridge University Press, 127–63

—— 2002. 'Theorising Health in the Context of Transition: The Dynamics of Perceived Morbidity among Women in Peri-urban Maharashtra, India', in *Medical Anthropology*, 21: 2, 157–207. Taylor and Francis

Layne, L., ed. 1999. *Transformative Motherhood: On Giving and Getting in a Consumer Culture*. New York: New York University Press

Lock, M. 2001. 'The Tempering of Medical Anthropology: Troubling Natural Categories'. *Medical Anthropology Quarterly*, 15: 4, 478–92. American Anthropological Association

—— and Kaufert, P. 1998. *Pragmatic Women and Body Politics*. Cambridge: Cambridge University Press

Lutz, C. and Lughod, L. 1990. *Language and the Politics of Emotion*. Cambridge: Cambridge University Press

—— and White, G. 1986. 'The Anthropology of Emotions'. *Annual review of Anthropology* 15, 405–36

Lyon, M. L. and Barbalet, J. M. 1994. 'Society's Body: Emotion and the "Somatisation of Social Theory"', in T. Csordas, ed. *Embodiment and Experience: The Existential Ground of Culture and Self*. Cambridge: Cambridge University Press

Martin, E. 1987. *The Woman in the Body: A Cultural Analysis of Reproduction*. Milton Keynes: Open University Press

Moore, H. 1999. 'Anthropological Theory at the Turn of the Century', in H. Moore, ed. *Anthropological Theory Today*. London: Polity, 1–24

Morsy, S. 1995. 'Deadly Reproduction among Egyptian Women: Maternal Mortality and the Medicalisation of Population Control', in F. Ginsburg and R. Rapp, eds. *Conceiving the New World Order*. Berkeley: University of California Press, 162–77

—— 1998. 'Not Only Women: Science as resistance in open door Egypt', in M. Lock and P. Kaufert, eds. *Pragmatic women and Body Politics*. Cambridge: Cambridge University Press

Parry, J. 1974. 'Egalitarian Values in a Hierarchical Society'. *South Asian Review*, 7: 2

Petchesky, R. 1987. 'Fetal Images: the Power of Visual Culture in the Politics of Reproduction', in M. Stanworth, ed. *Reproductive Technologies*, 57–80

—— 1995. 'The Body as Property: A Feminist Revision', in F. Ginsburg and R. Rapp, eds. *Conceiving the New World Order*. Berkeley: University of California Press, 387–407

—— 2000. 'Sexual Rights: Inventing a Concept, Mapping an International Practice', in R. Parker, R. M. Barbosa and P. Aggleton, eds. *Framing the Sexual Subject: The Politics of Gender, Sexuality and Power*. Berkeley: University of California Press

—— and Judd, K. 1998. *Negotiating Reproductive Rights: Women's Perspectives across Countries and Cultures*. London: Zed

Pigg, S. L. 1997. 'Authority in Translation: Finding, Knowing, Naming and Training "Traditional Birth Attendants" in Nepal', in R. Davis-Floyd and C. Sargent, eds. *Childbirth and Authoritative Knowledge: Cross-cultural Perspectives.* Berkeley: University of California Press, 233–63

Ram, K. 1999. 'Maternity and the Story of Enlightenment in the Colonies: Tamil Coastal Women, South India', in K. Ram and M. Jolly, eds. *Maternities and Modernities: Colonial and Postcolonial Experiences in Asia and the Pacific.* Cambridge: Cambridge University Press

Rao, M. 1999. *Disinvesting In Health: The World bank's Prescriptions for Health.* Delhi: Sage

Ragoné, H. 1995. *Surrogate Motherhood: Conception in the Heart.* Westview Press

—— 1999. 'The Gift of Life: Surrogate Motherhood, Gamete Donation and Constructions of Altruism', in L. Layne, ed. *Transformative Motherhood.* New York: New York University Press, 65–89

—— 2000. 'Of Likeness and Difference: How Race is being Transfigured by Gestational Surrogacy', in H. Ragone and F. Winddance Twine, eds. *Ideologies and Technologies of Motherhood*, 56–76

—— and Winddance Twine, F. 2000. *Ideologies and Technologies of Motherhood: Race, Class, Sexuality, Nationalism.* New York: Routledge

Rapp, R. 1999. *Testing Women, Testing the Fetus: The Social Impact of Amniocentesis in America.* New York: Routledge

—— 2001. 'Gender, Body, Biomedicine: How some feminist concerns dragged reproduction to the centre of social theory', in *Medical Anthropology Quarterly,* 15: 4, 466–78

Ravindran, T. K. and Panda, M. 2002. Gender and Fertility Transition in the South Asian Context. Paper presented at the conference on Gender and Health in South Asia. Heidelberg

Rozario, S. and Samuel, G. 2002. *Daughters of Hariti: Childbirth and Healing in South and Southeast Asia.* London: Routledge

Sen, G., Germain, A. and Chen, L. 1994. *Population Policies Reconsidered: Health, Empowerment and Rights.* Cambridge, Mass: Harvard University Press

Sheldon, S. 1997. *Beyond Control: Medical Power and the Abortion Law.* London: Pluto Press

—— 1999. 'Reconceiving Masculinity: Imagining Men's Reproductive Bodies in Law'. *Journal of Law and Society,* 26: 2, 129–49

Shore, C. and Wright, S. 1997. 'Policy: A New Field of Anthropology', in C. Shore and S. Wright, eds. *Anthropology of Policy: Critical Perspectives on Governance and Power.* London: Routledge, 3–34

Shweder, R. and Levine, R. 1984. *Culture Theory: Essays on Mind, Self and Emotion.* Cambridge: Cambridge University Press

—— 1994. *Thinking Through Cultures: Expeditions in Cultural Psychology.* Cambridge, Mass.: Harvard University Press

—— 2001. Deconstructing the Emotions for the Sake of Comparative Research. Prepared for Amsterdam Symposium on Feelings and Emotions

Stanworth, M. 1987. 'Introduction', in M. Stanworth, ed. *Reproductive technologies: Gender, Motherhood and Medicine.* Oxford: Polity Press

Strathern, M. 1992a. *Reproducing the Future: Anthropology, Kinship and the New Reproductive Technologies.* Manchester: Manchester University Press

—— 1992b. *After Nature: English Kinship in the Late Twentieth Century.* Cambridge: Cambridge University Press

Townsend, N. 1997. 'Reproduction in Anthropology and Demography', in D. Kertzer and T. Fricke, eds. *Anthropology and Demography.* Chicago: Chicago University Press

Unnithan-Kumar, M. 1997. *Identity, Gender and Poverty: new perspectives on Caste and Tribe in Rajasthan.* Oxford: Berghahn

—— 2001. 'Emotion, Agency and Access to Healthcare: women's experiences of reproduction in Jaipur', in S. Tremayne, ed. *Managing Reproductive Life: Cross-cultural themes in fertility and sexuality.* Oxford: Berghahn

—— 2002. 'Midwives among Others: Knowledges of Healing and the Politics of Emotions', in S. Rozario and G. Samuels, eds. *Daughters of Hariti: Birth and Female Healers in South and Southeast Asia.* London: Routledge

—— 2003a. 'Reproduction, Health, Rights: Connections and Disconnections', in R. Wilson and J. Mitchell, eds. *Human Rights In Global Perspective: Anthropological Studies of Rights, Claims and Entitlements.* London: Routledge

—— 2003b. 'Spirits of the Womb: Migration, Reproductive Choice and Healing in Rajasthan', in F. Osella and K. Gardner, eds. *Migration, Modernity and Social Transformation in South Asia,* Special Issue of Contributions to Indian Sociology. Delhi: Institute of Economic Growth

Van Hollen, C. 2002. 'Baby-friendly Hospitals and Bad Mothers: Manouvering Development in the PostPartum Period in Tamil Nadu, South India', in S. Rozario and G. Samuel, eds. *Daughters of Hariti: Childbirth and Healing in South and Southeast Asia.* London: Routledge, 163–82

Weiner, A. 1995. 'Reassessing Reproduction in Social Theory', in F. Ginsburg and R. Rapp, eds. *Conceiving the New World Order: The Global Politics of Reproduction.* Berkeley: University of California Press, 407–25

CHAPTER 1

ATTITUDES TO GENETIC DIAGNOSIS AND TO THE USE OF MEDICAL TECHNOLOGIES IN PREGNANCY: SOME BRITISH PAKISTANI PERSPECTIVES

Alison Shaw

Introduction

Clinical genetics in its modern form dates from the postwar period of the 1950s, and was initially a small hospital-based service provided by consultants, and the diagnosis of genetic disorders rested primarily on the observation of clinical features and on deductions about inheritance patterns based on the available family history. Laboratory techniques of genetic diagnosis were limited to chromosome karyotyping, which involves looking down a microscope to detect chromosomal abnormalities. Over the past fifteen years, however, molecular techniques for detecting submicroscopic genetic alterations have increased the number of identifiable syndromes, bringing a corresponding increase in the clinical workload, because the diagnosis of a genetic disorder has implications not only for the diagnosed person, but also for their parents, siblings, and own offspring.

A person with a dominantly inherited condition has a 50:50 chance of passing the condition on to a child, and, unless their parents have been assessed and declared free of the disorder (which would show it has arisen 'de novo' or as a new mutation), their own diagnosis also raises the possibility that a parent or sibling also has the disorder. A recessive disorder is one that arises when a child inherits two copies of a mutation, one from each parent; the diagnosis of a recessive condition therefore implies that *both* parents are 'carriers', that is, that they carry one copy of the mutation associated with the disorder. It also implies that any other children born to that couple also have a 1 in 4 chance of inheriting the disorder, and a 2 in 4 chance of being carriers themselves. Carriers of a recessive disorder are themselves perfectly healthy, but if they marry someone who carries the same mutation they have a 1 in 4 chance of having an affected child.

At least some of these implications are usually explained in the clinic when a diagnosis is made, or in subsequent counselling sessions, and clients are advised about the reproductive options open to them on the basis of their genetic risk. Today, an increasing range of disorders can now be detected in pregnancy, through antenatal screening and foetal anomaly tests (if available), and parents can choose to terminate an affected foetus. The philosophy of genetic counselling is that it should be nondirective, that counsellors should provide information and advice, but not direct clients' choices. Yet, as critics have observed, genetic counselling can never be entirely value free, because the very existence of the technologies that enable genetic diagnosis, the calculation of genetic risk and the detection of foetal anomalies implies that there is a decision to be made, either to continue with a pregnancy in the knowledge of risk or handicap, or to terminate where risk is high or a diagnosis is 'positive'. Such decisions are inevitably value laden (Rapp 2000). In relation to this, both participants and observers of the process of genetic counselling have expressed concerns that groups identified as 'at risk' of particular genetic disorders may be pressurised in subtle if not in overt ways to make particular kinds of reproductive choices, particularly if they are disadvantaged on the basis of social class or language.

Epidemiological evidence has shown an increased risk of recessively inherited genetic disorders among British Pakistanis (Bundey and Alam 1993). These include a wide range of malformations of development, including metabolic disorders, skeletal abnormalities, severe mental retardation and microcephaly. There has been much debate over the relative contribution of poverty and parental consanguinity in explaining adverse birth outcome among British Pakistanis (Ahmad 1996, Proctor and Smith 1997), for Pakistanis (and Bangladeshis) are the most disadvantaged of Britain's minorities in terms of government definitions of poverty (see Modood 1997). Nonetheless, the epidemiological evidence also raises issues of considerable social and clinical significance. There is concern that identifying Pakistanis as significantly 'at risk' of genetic disease will stigmatise an already vulnerable minority on the basis of its marriage pattern (Modell 1991, Ahmad 1996), and thus alienate potential users of the genetic service. Few Regional Genetics Centres in areas of substantial Pakistani settlement employ trained genetic counsellors with knowledge of Pakistani communities and of the relevant languages, Urdu and Panjabi (in its various regional dialects). In addition, relatively little empirical research has explored the issues that a genetic diagnosis and the provision of genetic risk information does in practice raise for Pakistani-origin users of the genetics services (see also Shaw 2000a, Shaw 2003).

Methodology and Research Participants

This chapter presents some preliminary findings from research involving clients of Pakistani origin referred to the clinical genetics department of a British teaching hospital.[1] The hospital is located in a town in which Pakistanis are the largest ethnic minority, comprising 3.7 percent of the population, and the local understanding is that most are from Mirpur district in Azad Kashmir. The data presented here are taken from informal interviews and participant observation with nine 'families'. 'Families' were identified through what geneticists call the 'index case' or 'proband': this is the child or adult about whom or for whom (in the case of a childhood or perinatal death or a termination of pregnancy) a diagnosis or a risk assessment is being sought. I have, when possible, met with the parent or parents, grandparents or other relatives (as appropriate) of the child, children or adult before their clinical interview, have sat in on their clinical appointments and have talked with them and with other family members at home on one or more occasions afterwards.

If a clinician, a paediatrician or a community-based general practitioner (GP or 'family doctor') suspects a genetic disorder in a child, they will arrange for the child to be seen in a genetics clinic. Four of the families discussed here were referred for this reason. If an obstetrician or paediatrician thinks that a malformation detected in pregnancy, recurrent miscarriage, or previous stillbirth, neonatal or infant death has a genetic cause, they will refer the mother to the genetics clinic. Four women were referred in this way, in one case following the termination of a pregnancy deemed 'incompatible with life'. Clients with a family history of genetic disorder may ask their GP to refer them to a genetics clinic for risk assessment: one of the couples discussed here requested referral for this reason. In theory, the issues raised by diagnosis are the same for all clients, in that a genetic diagnosis by definition implies certain recurrence risks. In practice, however, expectations of and reactions to the clinical consultation, may differ according to whether the referral is for a living child or children, for a woman or couple during pregnancy or soon after the experience of a termination or neonatal death, or for a couple with unaffected children as the case studies will indicate.

The chapter begins by introducing four cases of children referred to the genetics clinic. These cases illustrate a range of adult reactions to the idea or possibility of genetic disease (as opposed to disease of some other etiology). I then present some details of clients' accounts of their reactions to numerical risk information given by clinicians once a firm or provisional diagnosis is made, and of their subsequent reproductive decisions and experiences. For many of these families, religious considerations influenced attitudes to using reproductive technologies to prevent or monitor pregnancy or to determine a particular reproductive outcome. But at the same time, clients' accounts reveal a tension between theoretical statements about the use of such technologies and the realities of reproductive decision making. Religious considerations were also central in one couple's deliberations about using life-support techniques in the special care baby unit. These accounts reveal some of the practical ethical dilemmas that couples are facing, and indicate that solutions are being negotiated with respect to the authority of senior and other relatives, and the authority of religious leaders.

Case 1: Three Boys with Deafness

Riaz Ali is a 55-year-old unemployed former factory worker, from Mirpur district in Azad Kashmir. He has six children, three of whom are deaf and use hearing aids. He describes himself as uneducated, and he and his wife have poor English. When I visited them before the clinic appointment, Riaz Ali said he understood that the appointment was for the boys because of their deafness. He said that there was deafness in the family, because his wife is also deaf. He told me his wife's deafness was caused by a childhood accident, but he said he had no idea what had caused the boys' deafness: 'Why this deafness has happened, I don't know. Only God knows'. He believed the deafness might be cured by regular visits over a period of several weeks to his family *pir* (saint, religious specialist) in Pakistan. Unfortunately, though, he said, he could not afford the trip, and he would then have the additional problem of caring for his other children and wife. All of his children, he added, were behind at school.

At the clinic, the consultant made a genetic diagnosis, and told Riaz Ali that he had a 50:50 chance of having another deaf child. Riaz Ali said this does not matter because he is not planning to have more children. The consultant also advised that Riaz Ali's children might have deaf children. Riaz Ali commented that he knows a deaf boy and a deaf girl who married, and there is no deafness among their children. Ascertaining reproductive risks was not, for him, a current concern; far more pressing were his social problems, especially the fact that his children are behind at school. He described these concerns during the consultation, and the consultant asked if a letter explaining the diagnosis would be useful to him. He replied that although he cannot read, he would be able to show such a letter to doctors or teachers, to enable his children to obtain the help they need. Discussing the appointment afterwards, he said: 'All I want is help with my children's education. I want my children to be able to hear properly. I want them to study, to get on at school. I am not thinking of myself. My time is gone. But I want them to have a better life than I had'.

Case 2: Three Boys with Learning Difficulties

Mohammad Sarwar is a 57-year-old former factory worker, originally from Mirpur, with no formal education and poor spoken English. He has been unemployed for ten years. His wife also speaks very little English. The couple lives in a run-down council house on an estate where many of their neighbours are also from Mirpur. They have five children, three of whom are boys with learning difficulties, who attend special schools and had recently been referred to the genetic clinic. Before their clinic appointment, they told me that they had no idea why the referral had been made, and had not heard of 'genetics'. Nevertheless, Mohammad Sarwar intended to bring the boys to the clinic.

During the consultation, Sarwar spoke about one boy's constantly runny nose, and the eldest boy asked the consultant to arrange for plastic surgery to correct a leg injury. The consultant enquired about any family history of learning difficulties and medical problems. Sarwar said that no one else in his family had learning difficulties, and he did not think this could be a 'family' problem; he said his main worry was his wife's asthma, and he did not think there would be any value in exploring the possibility of a genetic explanation for the boys' learning difficulties. His own view was that his sons had become behind in the mainstream schools because the teachers were racist and had ignored the boys, while the boys themselves are lazy and pay little attention to what their teachers or their parents say to them. Taking into account Sarwar's views, and the fact that there was no obvious consanguinity in the family tree drawn up during the clinical consultation, the consultant concluded that there would be little point pursing any further the possibility of a genetic diagnosis.

Case 3: A 3-Year-Old with a Metabolic Disorder

Sohrab Hussain is a 3-year-old boy with a metabolic disorder involving obesity and severe learning difficulties. Before the clinic appointment, I met Sohrab's father, a young man raised in England, his mother, who has quite recently come from Pakistan, and his lively one-year-old brother. I also met his father's parents, who play a major role in caring for Sohrab, whose behaviour is very demanding and challenging. The family was very worried about Sohrab's behaviour, his lack of speech and his obesity. His mother told me she was afraid that his problems had been caused by medicines she had taken in pregnancy, while his grandmother told me she thought the problems might have been caused by Sohrab's circumcision operation, performed when he was a few weeks old. She and Sohrab's mother had not heard of the 'genetics' clinic, but they hoped that the doctor there would identify the cause of Sohrab's problems, and offer treatment. Sohrab's father was aware that geneticists identify illness that may be 'in the family', but, he emphasised, 'we have never seen this in our family before'.

Sohrab's parents took him to the first clinical consultation, and were told that their son probably has a recessively inherited disorder. His mother asked if the disorder could have been caused by the circumcision or by her taking medication during pregnancy, and the consultant assured her of no such link. The consultant said that the condition was most probably caused by a 'mistake' in the information about development contained in chromosomes, and a firm diagnosis should be possible in six months time, after Sohrab's chromosomes had been checked. Sohrab's father's main concern was with what the diagnosis implied for his son's development and health, and he was told that Sohrab will always be developmentally delayed, and will need regular medical checkups.

Case 4: An 11-Year-Old with a Metabolic Disorder

Nighat Sadiq is a college-educated woman of thirty-six, who came to Britain twelve years ago to marry her cousin, Nasir. They have three children. For them the idea of a genetic cause of their son's many medical problems is not new, because their second child was diagnosed with a recessive metabolic disorder soon after his birth eleven years ago. Over the years, Nighat and Nasir have attended numerous medical clinics, for their son's routine medical investigations and treatments. They were returning to the genetics clinic to explore the possibility of a more precise genetic characterisation of their son's condition. College-educated and from a major city in Pakistan, they had found information about their son's condition on the internet, were enquiring in the U.S. about the possibilities for research into their son's condition, and were hoping that the discovery of the gene that causes their son's problems would lead one day to a cure for him. To date, however, the gene remains unidentified.

The Novelty of 'Genetic' Disease

In Pakistan, rates of infant mortality are higher than in Britain. There, disabled and premature infants who might survive in Britain may be spontaneously aborted or die soon after birth, with no investigation of the possible genetic contribution to the outcome. At least in the areas of rural Panjab and Azad Kashmir from which most British Pakistanis originate, there is no directly comparable system for the genetic diagnosis of infant and childhood disability, or for the antenatal detection of birth anomalies. Beyond the major city hospitals, antenatal facilities and technologies for supporting premature births are scarce, and for city dwellers they are expensive. Nighat Sadiq commented that she was very lucky that Naveed was born in Britain, for in Pakistan they would not have been able to afford the healthcare he has received since birth. There, she added 'antenatal care costs 2,000 rupees a day, people earn 1,000 rupees a month, and only the very rich can afford health care, not people like us. And that is in Karachi'.

This may go some way towards explaining why, for all of the families discussed here in which a genetic disorder has been diagnosed, the condition appears to have arisen for the first time in the present generation in these families. In the clinical process of differential diagnosis, a clinician always asks if there is any known similar disorder in the family. One couple, who received genetic diagnoses for infants who died, speculated that the same disorder might have caused the deaths of infants born to a relative in the previous generation in Pakistan, but they had no way of confirming this. In all other cases, the answer is, 'We have never seen this thing before' or 'There is no-one else in the family'.

It is therefore not surprising that older parents such as Riaz Ali and Mohammad Sarwar, or grandparents such as Sohrab's, may consider these apparently new disorders to be linked in some way to the British environment. Sohrab's grandfather said it was strange that in Britain his son's first child was born with a disability, and doctors have told his son that this is a family problem, to do with something 'in the blood' and that his son risks having another affected child. 'In Pakistan', he said, 'my three brothers and my wife's brothers and sister each have between six and ten children, none of them disabled. There is more risk here than in Pakistan'. If the British environment has somehow caused the problem, then a return to Pakistan might offer a solution. Sohrab's grandmother told me that her younger son had also been obese, but she had taken him to Pakistan for three months and there he had lost several stones in weight. She thought that Sohrab's obesity and speech delay would likewise be corrected in Pakistan: he would lose weight and become 'smart'. She would also take him to a *pir* (spiritual leader) and then he would start to talk. She has already obtained a *tavis* (amulet) for him, containing words of the Qur'an, from a *pir* in Pakistan. Likewise, Riaz Ali thought that his son's deafness would be cured by regular visits to the family *pir* in Pakistan.

Beliefs in the power of *pir*s to cure terminal or incurable illness, including speech and other forms of developmental delay in children are central to the religious traditions of many Pakistani Muslims, but not necessarily shared by all, or with equal intensity (see Shaw 2000b: 204–12). Within a family, conflicting beliefs about illness causality and strategies of cure can generate tensions, particularly when only some members of the family have been formally involved in attending the hospital appointments. Sohrab's grandparents wanted to know what their son and his wife had been told at the hospital, but were unwilling to accept the diagnosis. Sohrab's grandmother cried a lot, saying she wanted to take Sohrab back to Pakistan and did not want to wait for the hospital diagnosis. She was very worried that Sohrab's blood and Sohrab's photograph had been taken at the hospital, and wanted to know why, and Sohrab's parents' explanation did not reassure her. Her reaction made them additionally anxious, but they, together with Sohrab's father's sister, who does most of the liaison with the special school Sohrab attends, felt strongly that they should wait until the medical investigations were complete before taking him to Pakistan. This uncertain situation continued, and it was only at a home visit from a genetic counsellor and interpreter some eight months later that Sohrab's grandparents had an opportunity to talk about the diagnosis and it implications more fully.

On the whole, older, pioneer-generation parents and grandparents are much less familiar with the terminology of chromosomes and genes than their sons and daughters raised in the U.K., but it does not necessarily follow that all younger adults are familiar with these concepts, especially since a large proportion of the spouses of second-generation Pakistanis are from Pakistan (Shaw 2001). One partner in all eight of the younger couples in the sample discussed here came to Britain relatively recently for or following marriage. One wife raised in the U.K. said of such husbands, 'they think like the older generation in these matters'. But other factors besides theoretical knowledge of genetics contribute to shaping reactions to diagnostic and risk information. Whether the couple has children already, the household's socio-economic circumstances and patterns of communication and patterns of authority between spouses and between other members of the household, as Sohrab's case illustrates, are also important, as well as prior experiences of having an affected child, or of infant deaths. Couples who are starting their families, or who have an unaffected child, may be initially skeptical about probabilistic risk information, but subsequent experiences of infant deaths may profoundly alter their perceptions.

Case 5: Two Infant Deaths

Farook Anwar is a 27-year-old college-educated computer programmer. His wife Farida is his first cousin, who came from Kashmir to Britain four years ago. They have one three-year-old son, who spent several weeks in the special care baby unit immediately after his birth, but is now well. They then had a daughter, who died in intensive care a few weeks after her birth. They were advised that this was a recessively inherited condition, and that they had a one-in-four chance of the same thing happening again. Anwar described his reactions at that time:

> I thought, what are they talking about? How do they know it? They said it is because we are first cousins, but there had not been anything like this in our family before. Plenty of people marry cousins without having handicapped children, and some people have handicapped children without marrying cousins. So we decided to go for it. Perhaps it was selfish, but we wanted our son to have a brother or sister to play with.

Their second affected daughter died in intensive care after five weeks. Shortly after her death, Anwar told me: 'After this, we do not want to have more children; not for 5 or 6 years at any rate. We have been today for birth control'. His wife was lying on the sofa at the time, and scarcely spoke throughout the interview. Anwar remarked: 'She is very upset. She feels she cannot produce healthy children. She cannot perform the function for which she is here. I do not blame her. It is a cultural thing that she feels like this'.

Case 6: Three Infant Deaths

Liaqat Ali is a 25-year-old medical technician. His wife is his cousin, who arrived from Pakistan five years ago. They have one child, a healthy daughter. A second daughter was diagnosed soon after birth with a rare recessive condition, and died after three months. A year later they had a son, who was also affected, and lived for five months. Their fourth child was born with the same disorder, and lived for seven months. Having experienced three infant deaths from this disorder, Liaqat Ali said: 'I want to go back to Pakistan for a year or two. We are not going to have any more children, that's it. I feel we have really been dragged through the mill. It is hard staying in the same house and going to the same hospital and pretending nothing has happened'. His wife returned to Pakistan for a year. He wonders, in the light of such bad luck, if the odds might now be more heavily stacked against them than with the standard recessive recurrence risk, for reasons that medical science may, or may not, be able to explain. They have requested further genetic counselling, to review their risk assessment.

Liaqat Ali was advised of the recessive risk when their first affected child was diagnosed, although he did not explain it to his wife, who speaks no English. They both understood, however, that there was a risk of the same thing happening again. Liaqat Ali recalled:

> It was a worrying next pregnancy, as there were no tests available. A week before the birth, I asked the gynaecologist, what is the chance of this happening again? He said, 'it is like me telling you I am going to the moon and back'. He lived six weeks longer than his sister; probably because he was diagnosed earlier and had the right medicines when he needed them, but it was the same. The paediatrician said we had been very unlucky, that there is the risk element that it will happen again, but it is okay to try for a baby, the risk is no greater than before. He said, 'It is unlikely the dice will land again on a six and a six'. We thought our bad luck was over and the next child will be okey. But if we had known, we would have stopped, for three or four years at least. All that time, we were trying to fill that vacuum up, to replace the lost baby. But instead, the vacuum got bigger and bigger. In the probability sense, the chances should be less and less, we thought it would be less and less, when the next baby was born. But it was not. The girl to who did the EEG (electroencephalogram) for this baby was the same one who did it from the other two. I could see from her face; I did not need to wait for her to tell us the result.'

Interpreting Risk Information

One consequence of these painful experiences of infant death or childhood disability, in the short-term at least, is that couples may decide to have no more children, or at least to postpone further pregnancies. Liaqat Ali's wife returned to Pakistan for a year, and Mohammad Anwar insisted that, in the immediate future at least, the couple prevent any further pregnancies. These reactions can generate disagreement between couples and between generations; they can cause much distress for young wives for whom having children is often of central importance to strengthening their positions within their husband's family. But on the other hand, women may be more insistent than their husbands that they do not want, yet, to risk another affected child. Nighat Sadiq recalled, after her second child was diagnosed with severe physical and learning disabilities that would require a lifetime of medical surveillance, that

> I had no intention of having another baby, because I have high chances. That is one reason for our slow time to have another baby: nine years was a long time. We just did not want to think about it. And it is not fair to the child either, is it. He is suffering at the moment. If you can avoid it, then it is better. I mean, that's how we look at it.

Recurrence risks may sometimes be inflated in recollections of this time, perhaps reinforcing parents' desires not to repeat the emotional stress of their experiences. Discussing the recurrence risk they were given, Nasir recalled, 'We were told that there is a one in four chance of having another child like this'. Nighat, however, contradicted him, 'Three out of four', she interrupted. 'No, one out of four', he corrected. 'The point is', Nasir went on, 'we did not have another child for a long time after that. We were too frightened the same thing would happen again, and they would not be able to do anything'.

Similarly, during the evening in which I spoke to him shortly after the second infant death, Anwar gave an inflated recessive risk when telling me about the information they had received after the first infant death: 'We were told that we had a three in four chance that the next baby would have similar problems. A three in four chance is high, but people don't like to think about that, they turn a deaf ear, and I did not take it seriously'. In fact, he was given a standard recurrence risk of one in four, which gives a three in four chance of having an unaffected child.

Another difficulty with interpreting probabilistic risk information is that people may understand a one in four risk to means that one child out of four will be affected, so that if they already have had an affected child, their risks will be lessened in subsequent pregnancies, whereas in fact the risk remains the same with each pregnancy. Liaqat Ali feels that the risk information they received gave them a false sense of hope. At first, they had understood, this was a 'one off' genetic disorder, so they tried for another baby. He may have understood the disorder to be 'one off' because of the stated recurrence risk of one in four. The fact that this risk was described as being each time 'no greater than before', may have led him to conclude that the chances of an another affected child would diminish each time an affected infant was born, as if their bad luck was being used up, 'We thought our bad luck was over and the next child will be okay ... In the probability sense, the chances should be less and less, we thought it would be less and less'.

Case 7: Risk of a Chromosomal Anomaly

Hamid Ullah is a 28-year-old computer programmer, whose wife Humairah has been in England for three years. They have 1-year-old twin boys, but Hamid has two brothers with a genetic disorder linked to a chromosomal rearrangement. Tests have shown Hamid carries a balanced form of the chromosomal rearrangement linked with his brothers' problems, and there is a risk that children he has will inherit the unbalanced form, or inherit the same risk as his. Hamid and Humairah attended the genetics clinic to clarify these risks. Shortly beforehand, a health visitor told me that Hamid Ullah is 'very religious' because he refused amniocentesis during his wife's pregnancy. During the clinical consultation, a registrar commented, 'I know it is against your religious beliefs to have a test during pregnancy'. Hamid corrected her, saying it was because of his personal rather than religious belief that he was against this test: 'The baby is there, isn't it? The test could kill it; they said so'. After the appointment, Humairah told me that she would not want any test that might result in her losing the baby.

Case 8: Infant Death, Recessive Risk and Use of Amniocentesis

Haroon Modood and Shabnam Bi's first child was diagnosed at three months after her birth with a rare recessive disorder. They understood that 'this has come from our blood because we are cousins. We have a one in four chance of having another child like this', and accepted a foetal anomaly test during the next pregnancy. 'They took a piece of our daughter's skin, and some liquid from my wife's stomach, and checked that the baby did not have the same thing that is in our daughter's skin'. The baby was unaffected, but I did ask if they would have considered termination had the result been 'positive'. Haroon Modood looked at me blankly, asked me to repeat my question. After a few minutes reflection said: 'No, I don't think so. No, I wouldn't want to do that. We just wanted to know'.

Monitoring Pregnancies

Women or couples with a known risk of having an affected child are likely to be offered ultrasound scans in any subsequent pregnancies, and may be offered foetal anomaly tests if these are available for the disorder in question and, by implication, the opportunity to terminate an affected pregnancy. From a clinical viewpoint, antenatal tests are offered in order to give women the opportunity of terminating an affected child. Clinicians sometimes assume that Pakistani Muslim clients will reject all prenatal testing on religious grounds, because to accept prenatal testing implies the possibility of terminating pregnancy, which, they believe is unacceptable in Islam. As Hamid and Humairah's case illustrates, a reputation for being 'very religious', and therefore opposed to all antenatal tests, may sometimes precede a couple to their clinic appointment. Many Pakistani Muslim clients will say that termination is unacceptable in their religion, and tantamount to murder, a sin (*gunah*) for which God will punish you. In practice, however, the relationship between religious belief and the realities of reproductive decision making are more complex.

In the first place, my interview data suggest that clients' frequently separate the process of seeking information from that of making decisions about what should be done when information is obtained. All of the women or couples I have spoken with consider that ultrasound scanning is acceptable because, they say, it does not harm the baby. Humairah, who was opposed to amniocentesis because of the risk to the baby, was quite open to the idea of using ultrasound technology to provide information. If the scan shows a baby to be handicapped, she said:

> You cannot do anything. It is in the hands of God. You cannot 'waste the baby' (terminate), because when the chromosomes from the mother and

from the father mix together, the baby starts, and there is a soul. You cannot kill the soul. That is *gunah* (sin), and God will punish you.

So what, I asked her, is the advantage of having a test if you cannot do anything? I expected she would talk about preparing for the birth of a child with a disability, but instead, she replied:

> The other night, my brother-in-law was listening to the news, about the weather. They said there would be two hot days, and then a storm. There were three hot days, and there was no storm. Science can tell you a lot of things, but they cannot tell you everything, and sometimes they get it wrong. Only God knows everything, and he decides how it will be.

She went on to expand on her view that scientific knowledge is inherently limited:

> In Pakistan, a couple decided they wanted to stop more babies coming. They already had children. So the woman had an operation. After that, they had two more children, and both were handicapped. It shows, God decides about your children, how many you have, and how they are, whatever you do.

This does not mean that parents do not worry about the risks of having disabled infants. Hamid might confidently defend his personal belief to potential critics, but his wife says he worries a lot, 'because there is a big difference between having a child that is handicapped and having a child that is not. But he also says, and in our Islam it is like this, that the kind of child you have is in the hands of God.'

Despite the risk of spontaneous abortion, women and couples do sometimes agree to amniocentesis. Harood Modood and Shabnam Bi had an amniocentesis test, but said they would not terminate an affected child; whether or not they would have opted for termination had the test shown the child was affected, I cannot say. Termination is not easily talked about or admitted to. But women at risk of having another child with a disability detectable by amniocentesis do sometimes choose to have the test, and to have abortions if the test identifies the disorder, sometimes without telling their husbands or even closest relatives, for fear of the disapproval of family and community.

Two local *imams* explained to me that the pursuit of scientific knowledge is allowed in Islam, and that Islam permits termination within the first three or four months of pregnancy (also see Shaw 2003). Few of the women I spoke to were very clear about the conditions under which Islam might permit termination. Nighat Sadiq, however, considered that Islam permitted termination in early pregnancy. Her decision to opt for termination, if a test could confirm the presence of another affected foetus, was based primarily on the fact that she felt she would not be able to cope with another affected child. She had become unexpectedly pregnant eight years after deciding not to risk a further pregnancy, but was very distressed to discover that that a test was not available for the rare recessive condition in question:

> When I knew I was pregnant, at first I was very depressed, because I had been through it all with Naveed. The feedings problems, everything. I was so scared the same thing would happen again. They had told me to tell them if we planned another baby, so I went to the consultant, and she offered me an early test, where they take something from the womb. I can't remember what the test is called. But then they changed their mind, because, they said, they would not be able to tell me for sure if the baby is affected or not. They said they could not tell me until the very last, not until after four or five months, if the baby would have any problem. I was very upset by that. I mean, they told me I could have some tests when I was four months pregnant, and if I want, you know, to lose the baby after that (she is speaking hesitantly now, and uses the Urdu idiom for termination: *zaia karna*, literally to lose or waste the baby), but it was a very late stage. I just did not want. There is no point after four months ... I wouldn't mind going for the very early stages.

Case 9: Termination of a Pregnancy Deemed 'Incompatible with Life'

Kaniz is a 26-year-old marketing trainee raised in the U.K. whose husband, a first cousin, has been in England for two years. Scans during her first pregnancy showed a baby that would be stillborn or die immediately after birth because of multiple structural abnormalities. She decided to terminate. Knowing that she has a three in four chance of an unaffected child, she plans to have another child in due course, but not for a while yet, and has returned to fulltime work. Her husband, however, would like another baby straight away, to 'fill the gap'.

In this case, Kaniz decided to terminate the pregnancy mainly through discussion with her sister. Their reasoning hinged upon a distinction between a baby that would die at birth, and a baby that would have a life as a handicapped child. Her sister explained:

> My sister carried her first baby for six months but the doctors said the baby would die within an hour or two of being born, because they found many things wrong inside him. They said he might live only half an hour. It was not a 50:50 situation. He would certainly die. His heart was wrong, his liver was wrong. His kidneys were very large. His stomach was full with kidney. He would not live for more than a couple of hours. My sister thought it over, and her parents, and the family. We did not want her to carry it to term. The family all agreed it was not fair on her to carry for nine months a baby that would die. It is not as if he is a handicapped baby. If God gives you a handicapped baby, you would still love them, and look after them. But this baby was going to die. My sister said she would think it over.

In fact, her sister's husband was not convinced that termination was the right decision:

> The next morning my sister rang me to say that her husband won't let her go through with the termination. He wants to keep the pregnancy. He is like

our parents' generation in his thinking. He came from Pakistan for marriage. He did not go to school here. He thought the doctors had made a mistake. He thought, God knows, they don't know anything. How can they say? Then I spoke to him, and explained. And then he agreed.

However, three months later, the parents were still grieving, and Kaniz's sister felt responsible, and unsure that she had made the right decision: 'I made her make the termination. I don't know if I made the right decision. Someone told me the other day that they had been told they had a high risk of an abnormal baby, but they kept it, and the baby turned out to be okay'.

Withdrawing Life Support

Farook and Farida's second baby (case 5) was shown on scans to be five weeks too small for the dates, so Farida was kept in hospital for the rest of her pregnancy. She gave birth by Caesarian section to a baby weighing 2.5 pounds who was taken straight into intensive care in the SCBU (special care baby unit): 'There were more tubes and medicines than baby, she was so small ... They did so many tests, and she was in there for two months; more than two months of tubes and medicines to keep her going. Everything in her body was not working properly ...'

Anwar was then asked to decide if medication should continue, and if the baby should be kept alive with the assistance of a ventilator:

> I felt they wanted an answer because they wanted the bed for another patient, but it was very hard to decide. I thought we cannot take the tubes and medicines away, that is like murder, she will die without them, so I said they must keep that going. They must give her the chance. If she has the strength, if she has the will, she will live.

But about the ventilator, he was uncertain:

> When I looked at her, I felt inside me that it would be wrong to put her through even more suffering by doing more. If she survives, what kind of life will it be, if she is so dependent? She cannot do anything. She is more tube and medicine than baby. What extra suffering would that bring her?

Towards resolving the dilemma, he consulted his *pir*, whose ancestors had offered spiritual protection to his family in Kashmir for generations, and who happened to be visiting Britain at the time:

> He said, in any religion, you are told that you must not take a life. It is the same in Christianity. Thou shalt not kill. It would be murder to give her an injection, to help her die. You cannot do that. I took him to see her. I thought, he should see for himself what she is suffering. Often people say things to a *pir*, making them up, just to get an opinion that suits them. I wanted him to understand. He stayed there about half an hour. I think he

was disturbed by what he saw. He said he thought it would be wrong to make her die, but it would be wrong to increase her suffering. So I said to the doctors, just keep the tubes and medicines, but don't do anything more, no ventilator now. It is up to her, we will see if she has the will to live or not. And I could see, with her mouth open, no control, that it was too much for her.

He felt that he could now live with his decision, knowing that the doctors had done what they could, that they had given his daughter a chance, and because his *pir* had reassured him that 'I have not done anything that could be considered murder, which would be a sin in my religion, and in any religion'.

Conclusions

British Pakistanis have been identified to be at particular risk of recessively inherited disease, in comparison with other ethnic groups, and this has been linked to the practice of consanguineous marriage, a finding that raises sensitive issues of policy. However, little is known about their attitudes to genetic disease and to genetic risk information. This chapter has demonstrated a spectrum of engagement with the genetic diagnostic and antenatal services, and how this engagement is negotiated with reference to a wider non-medical framework, of family and religious community. Some families are referred with little or no idea of what a genetics service might offer, drawn into the system through the process of clinical referral. In all cases discussed here, the idea of a 'genetic' illness is new, in that, although families may be aware of disorders 'of the blood' and 'in the family', the syndromes diagnosed seemed to have arisen for the first time. This may, in part at least, be explained by the fact that in Pakistan, infants with major birth anomalies are unlikely to survive birth or live as long as can occur in the U.K.

For families with affected children, usually the first concern is to find appropriate treatments and cures. Treatments may involve the use of 'folk' remedies and religious specialists and taking the child to Pakistan as well as making use of western biomedicine, and relatives with different levels of involvement with caring for a child may disagree about the best course of action. Senior relatives, such as grandparents and parents-in-law, may play an important part in household decision-making in these matters, but may not have attended the child's clinical appointments, so that younger adults then have to negotiate the implications of clinical results with other family members. In this process, conflicts may arise over the implications of diagnosis for a particular child and in subsequent pregnancies. Attitudes to the use of medical technologies in pregnancy are also often ambivalent. There is support for the idea of pursuing scientific knowledge, but its results are sometimes to be treated with a degree of skepticism, and there is concern about the risk of harming a foetus by using particular antenatal tests. Termination is viewed even more ambivalently, for it is generally believed to be sinful. Women who have terminations may feel grief and guilt, tending to discuss the matter, if they discuss it at all, only with their closest relatives or friends. Women or couples in these families are negotiating new ethical territory, sometimes drawing on religious leaders for advice where the way ahead is unclear.

These findings raise issues relevant to antenatal and genetic counselling provision, in general, and not only to the issue of recessively inherited conditions. In relation to the incidence of recessive disorders among British Pakistanis, the question of whether the pattern of consanguineous marriage will continue is also of clinical relevance, and the findings presented here, while not directly concerned with this question, have some implications for it. Among some families, there is already defensiveness about cousin marriage, where the medical view is taken to be that the disorder has arisen 'because we are cousins'. It will not necessarily follow from this that marriage within the family will cease. While marriage plans for their children are not, in most cases, uppermost in the minds of parents struggling to care for handicapped children or trying to give birth to a healthy living child, in some cases parent have siblings or older children approaching marriage and parenthood who are asking questions about their own risks of having affected children. At least for some families who have personal experiences of infant death and disability, it is likely that genetic risk information is likely to be taken into account along with other factors that influence marriage choices and processes of marriage arrangement.

Note

1. I am grateful to the Wellcome Trust for a project grant that supports this research. I wish also to thank the clinicians and research participants who have made this project possible. All names have been changed.

References

Ahmad, W. 1996. 'Consanguinity and related demons: science and racism in the debate on birth outcome', in C. Samson and N. South, eds. *The social construction of social policy*. London: Macmillan, 69–87

Bundey, S. and Alam, H. 1993. 'A five-year prospective study of the health of children in different ethnic groups, with particular reference to the effect of inbreeding'. *European Journal of Human Genetics*, 1, 206–19

Modell, B. 1991. 'Social and Genetic Implications of Customary Consanguineous Marriage among British Pakistanis'. *Journal of Medical Genetics*, 28, 720–23

Modood, T., Berthoud, R., Lakey, J., Nazroo, J., Smith, P., Virdee, S. and Beishon, S. 1997. *Ethnic Minorities in Britain: Diversity and Disadvantage*. London: Policy Studies Institute

Proctor, S. R. and Smith, I. J. 1997. 'Factors associated with birth outcome in Bradford Pakistanis', in A. Clarke, and E. Parsons, eds. *Culture, Kinship and Genes*. London: Macmillan, 97–110

Rapp, R. 2000. *Testing Women, Testing the Fetus: the Social Impact of Amniocentesis in America*. New York and London: Routledge

Shaw, A. 2000(a). 'Conflicting models of risk: clinical genetics and British Pakistanis', in Pat Caplan, ed. *Risk Revisited*. London: Pluto Press, 85–107

—— 2000(b). *Kinship and Continuity: Pakistani Families in Britain*. London: Routledge/Harwood Academic Publishers

—— 2001. 'Kinship, Cultural Preference and Immigration: Consanguineous marriage among British Pakistanis'. *Journal of the Royal Anthropological Institute*, 7, 315–34

—— 2003. 'Genetic counseling for Muslim families of Pakistani and Bangladeshi origin in Britain, The Encyclopedia of the Human Genome'. London: Nature Publishing Group, 2, 762–66

CHAPTER 2

LOCALISING A BRAVE NEW WORLD: NEW REPRODUCTIVE TECHNOLOGIES AND THE POLITICS OF FERTILITY IN CONTEMPORARY SRI LANKA

Bob Simpson

Recent advances in embryology, molecular genetics and related clinical fields have brought about rapid and far-reaching changes in reproductive medicine. The proliferation of techniques for the collection, screening, storage and implantation of egg, sperm and embryo significantly extend the possibilities of how conception and gestation might take place and who may be involved at the different stages. Uncoupling heterosexual intercourse from conception and, in turn, conception from corporeal gestation opens up new possibilities for the way that kinship, reproduction, gender and the state articulate one with another. Social scientists have contributed significantly to debates about the ethical, legal and social impact of the new technologies through their attempts to locate what might otherwise be seen as narrow issues of technological development in medicine, within broader social and cultural frameworks.[1] Not surprisingly, attention has focused mainly on western societies, that is, on the societies where these technologies originated and have proliferated as part of the bigger project of medicalising and regulating human reproduction. However, it would be a mistake to assume that the development and impact of the new technologies are only felt in the first world. The development of new medical knowledge, expertise and technology is not confined to Europe, Australia and North America but is increasingly an aspect of the medical systems of less developed countries. Novel procedures involving the manipulation, storage and transportation of eggs, sperm and embryos are now available across much of South Asia. The impetus for the expansion of fertility services has been driven by the prospect of remedying the deep distress and stigma faced by childless couples (Riessman 2000). Indian clinicians were reported to have produced their first baby by in vitro fertilisation (IVF) baby as early as 1978 (Bharadwaj 2000, 2002) and there followed a spectacular growth in the use of techniques such as IVF and intra-cytoplasmic sperm injection (ICSI).

Sri Lankans have similar concerns over fertility and the remedies that might be put in place should it be impaired in any way. Over the last decade a wide range of developments have taken place which have brought the new reproductive technologies to bear on the problems of a few and into the consciousness of many. The transfers of technology and ideas which underpin such developments are not uni-directional but, as Ginsberg and Rapp (1995: 1) point out; 'People everywhere actively use their local cultural logics and social relations to incorporate, revise, or resist the influence of seemingly distant political and economic forces' (see also Lock et al. 2000). Furthermore, the process is not without conflict and, as in the West, involves what Rabinow (1999: 12) refers to as 'contestations over how technologies of [social and bodily] combination are to be re-aligned with technologies of signification'. Beyond the paradigms of Euro-American kinship, however, 'contestations' over what can be brought together in the name of procreation invoke different ideas of kinship; different ways, that is, of understanding how persons see themselves as connected by substances and sentiments to one another, to their past and to their future.

Some of these issues were explored in a short study carried out in Colombo, Sri Lanka among those involved with the reception and promotion of the new technologies.[2] Part of the research involved discussions with doctors, academics, clinicians and the members of several local medical ethics committees about the new technologies and their reception in Sri Lanka. In considering these various specialist and lay accounts I was struck by the way that, in a time of profound social and political uncertainty, connections were made between the new and uncertain realms of the new reproductive and genetic technologies on the one hand and rather more familiar and certain worlds of custom and practice on the other. As the majority of my informants were Sinhalese and Buddhist these responses were drawn from a distinctive cultural repertoire. They demonstrate how scientific and technological developments are cast in local idioms. References to historical patterns of kin relations as well as more contemporary reflections on abortion, organ/blood donation and adoption frequently surfaced in discussions as the logical reference points used to navigate a way through the new and unfamiliar debates in which they were being asked to participate. My aim in this chapter is to bring these references together in order to throw light on the changing politics of fertility and reproduction in contemporary Sri Lanka.

Reproducing the State and the State of Reproduction

At a Buddhist ceremony in June 2001, Sri Lanka's then Prime Minister, Ratnasiri Wickremanayake, called for citizens to abandon the country's highly successful practice of family limitation and start having more children. The Prime Minister urged people to realise that, where families are concerned, 'big is better'. Needless to say the comments caused great controversy, not least among family planning specialists who had long campaigned under the slogan 'small is beautiful'. The existing policies had been seen by the United Nations Population Fund as a great success and hailed as something of a model in the region.[3] In 1999, the population was approximately nineteen million people with growth at an estimated 1.4 percent. Life expectancy was in the region of 72.5 years and infant mortality at fourteen per thousand births. However, the Prime Minister was not so much interested in model family planning policies but giving voice to an altogether different set of anxieties. The prompt for the Prime Minister's comments was the failure of a very public campaign to enlist 10,000 soldiers to fight in the war against Tamil secessionists and 2,000 Buddhist monks to join the priesthood (*sangha*). Few men came forward to take up either the gun or the robe; the inability to replenish the ranks of two of the key institutions committed to preserving Sinhala Buddhist identity and hegemony was a troubling sign of the times. Significantly, the failure of the campaign was not seen as an expression of disenchantment with the way in which the government had handled the war, the economy or its relations with the Buddhist clergy. Rather, the blame was laid with families in general and women in particular who were failing to do their bit, to create sons who would heroically preserve the Sinhala Buddhist heritage by becoming 'world conquerors' and 'world renouncers' (Tambiah 1976).

In calling for a drive to increase the population, Prime Minister Wickremanayake was giving expression to the fear (*baya*) felt by many Sinhalese about the future of the island in the light of its deeply troubled recent history. Although development has continued throughout the various upheavals of the last twenty years it has been seriously uneven and largely centred on the capital, Colombo. The main reason for this unevenness has been a long running conflict between the predominant Sinhala Buddhist community (approximately 74 percent of the population) and the Tamil Hindu community (approximately 18 percent) with elements of the latter engaged in a bitter war for an independent homeland in the north of the island. Colombo has been the site of numerous bomb outrages which have impacted very badly on the mainstays of the Sri Lankan economy, namely the financial and tourist sectors. The south of the island has also experienced its own violent convulsions. In 1971 and 1989 insurrections led by educated but disenchanted rural Sinhala Buddhist youth, resulted in widespread civil disorder. Estimates vary but as many as 100,000 people may have been killed as a result of ethnic and civil strife in Sri Lanka over the last two decades.[4] In short, the image of the smiling, carefree Sri Lankans who feature in the tourist brochures has proved difficult to sustain in the face of such massive distress. Low wages, high inflation, an expensive war of attrition, chronic traffic congestion and pollution, cynicism about politicians and the democratic process, corruption, the rise in crime and violence are but a few of the things that make up the mix of tribulation which is nowadays the daily lot of most Sri Lankans living in the capital.

The fear expressed, however, is not just about whether a routine trip to the office or the shops will be one's last but it is also a more abstract concern about the future. Will the island succumb to partition? Will the democratic process finally disintegrate into anarchy? Will there be another insurgency in the south and this time will it be successful? Can the present institutions of civil society adequately contain the island's ethnic and cultural pluralism? In short, the nation state, particularly as it is conceived of by Sinhala Buddhists, is felt to be under threat; attempts to reproduce the values, traditions and institutions associated with it to date may yet fail; there is a sense in which at the turn of the millennium Sri Lanka finds itself at a crossroads (see Hettige and Mayer 2000). Thus, at a time when Sinhala Buddhist hegemony is arguably as strong as it has ever been, there is a growing perception that the Sinhalese and their distinctive tradition of Theravada Buddhism are more vulnerable than ever. As one leading monk put it in an interview in 1987, the island is like a 'tiny ... fragile Sinhala Buddhist society ... a tear drop, a grain of sand, in an enormous sea' (Juergensmeyer 1990: 58). Similarly, fears about current vulnerability and future extinction are played upon in the popular press as well as in the literature of Sinhala Buddhist nationalist groups such as *Sinhala Urumaya* (literally Sinhala Heritage). But, crises evident in the reproduction of the nation state also have powerful resonances when it comes to the problems encountered in the reproduction of individual people. As Heng and Devan (1992: 349) suggest, there is a fundamental link between 'reproducing power and the power to reproduce'. Each can become a metaphor for the other, a means to think about the future and how one's connections with that future may be made possible or denied. It is not surprising, therefore, that the bouts of acute postcolonial anxiety to which Sri Lanka is currently prone manifest themselves in concerns about the family in general and the role of women in particular. In recent years, the rise in frequency of child abuse, mass migration of women to the Middle-East, awareness of endemic domestic violence and the demise of the extended family have become the focus of government and popular concern, for they appear to strike at values and relationships upon which Sinhalese family life is founded. There is deep apprehension that distortion of traditional arrangements for family, residence and reproduction will in various ways imperil the country's future. Likewise, concerns about circumstances which are having a direct impact upon the fertility of men and women have also attracted considerable amounts of attention. Perhaps the two most controversial in this regard are the regularity with which women resort to terminations when they experience an unwanted pregnancy, on the one hand, and the decline of male fertility on the other.

Reproduction Imperilled

Buddhists are committed to an ethic of non-killing and abortion is an abomination which attracts deep opprobrium. As Harvey (2000: 314) points out, the intentional killing of a child in the womb is particularly emotive because 'for a being to gain a foothold in a human womb and then be killed is to have this rare opportunity destroyed'. It is not surprising to learn therefore that abortion is strictly illegal in Sri Lanka; it is only available where a mother's health is directly threatened and three physicians are prepared to sign a form to this effect. What is surprising to learn, however, is the high rate of terminations that take place on a daily basis. There are no official figures for the current abortion rate but unofficial estimates put the number of terminations in the region of 700 per day (De Soysa 2000: 46). The majority of terminations appear to be sought by older married women for whom contraception has failed. Amongst these women there is a high rate of death (estimated to be between 10–30 percent of all maternal deaths) due to sceptic abortion (Ibid.). The legislation that allows and indirectly promotes this state of affairs is seen by women's groups and large sectors of the medical profession as regressive and oppressive (Abeyesekera 1997). Even though the law on abortion was changed in 1995, reforms stopped far short of what many campaigners wanted. Attempts to modify the penal code so that abortion was available in cases of rape, to unmarried mothers, cases of incest or where there are foetal abnormalities were unsuccessful. Many of those involved in campaigns to change the law on abortion were of the view that the government was more concerned not to offend powerful religious groups in the country and particularly the Buddhist clergy (*sangha*), who remain implacably opposed to abortion, rather than to address the chronic problem of maternal health caused by back-street abortion. Furthermore, some interests in the abortion debate were not just about the morality of taking life but resonate with broader concerns about threats posed to the reproduction of Sinhala Buddhist cultural identity. There is a widely held view that when it comes to abortion and contraception, the Sinhalese are, in practice, far more liberal than their neighbours from other ethnic groups. Many Buddhists believe that effective family planning is differentially practiced, with the result that other groups such as Hindus and Muslims were outstripping them in terms of population growth. One experienced general practitioner suggested that in his experience some doctors put women seeking effective contraception (e.g., tubal ligation) or terminations under considerable pressure to continue bearing children. For some members of the Sinhala Buddhist community, it would seem the demographic tide is believed not to be flowing in their favour and the only way to stem it is to ensure that mothers produce sons and daughters in greater numbers than they are doing at present.

Anxieties over the threats that abortion causes to rates of reproduction are matched by parallel concerns over male reproduction. Over the last two decades the male death rate has massively outstripped that of females. The discrepancy in the age range 15–34 is particularly pronounced due to the effects of the war, social violence, road traffic accidents, suicides and poor occupational safety (De Soysa 2000: 41). Real demographic concerns over the demise of Sri Lanka's finest young men is compounded by the loss of reproductive potential of another kind. In traditional Sinhalese medicine semen (*sukra dhātu*) is considered the highest of substances. Through a process of 'cooking' using the heat of the body, food is converted into ever finer portions (*āhāra prasāda*) and the highest of these is semen; one drop of which is believed to equal to sixty drops of blood (Obeyesekere 1976: 201). Although beliefs about the mental and physical consequences of semen loss may not be quite as prominent as they were in former times, semen remains a powerful symbol of male reproductive potency and what happens to it, inside or outside the bodies of men is a cause for concern. Whilst carrying out fieldwork, several newspapers carried reports that male infertility is on the increase due to agricultural chemicals. Increasingly, it would seem, sperm are not strong enough to make the 'long difficult journey' (*dirgha gamanaka yēdi*) to meet the egg (*dimba*) (cf. Martin 1991). Late marriage, smoking and use of drugs are believed to contribute to increased levels of testicular cancer, small testicles and the absence of sperm which leave many men is a state of mental stress (*mānasika ātatiyen*). At the same time, newspapers were also reporting a thriving black market in the sale of Viagra and growing concerns about how to regulate its import and distribution. A variation on these themes was provided by a colleague carrying out fieldwork in a remote part of the south-east of the island. She reported how two of her informants were convinced that the increase in male infertility was an international plot to eliminate the Sinhalese. The means to achieve this end was the German measles vaccine which at the time was being administered across the country by various NGOs (pers. comm. Mariella Marzano).

Anxiety that far too many babies are being aborted at a time when male reproductive potency is in decline engenders popular fears that collective reproduction is becoming fragile and endangered. Concerns about infertility, however, are nothing new. In Sri Lanka, as in all South Asian societies, sub-fertility is a deeply distressing condition. A high premium on family ties, the importance of their extension through procreation for purposes of inheritance and the need for children as welfare in old age mean that childlessness does not go unremarked. Couples are placed under considerable pressure from family members (and particularly from the husband's mother) to produce children shortly after marriage. Where this does not take place women usually bear the brunt, stigmatised and perhaps considered inauspicious in later life. Traditionally, couples have resorted to a wide range of healing and diagnostic practices, such as making offerings at the temple, petitioning the gods (especially *Pattini*), visiting astrologers (to identify planetary influences on fertility) or exorcists (who might perform the *rata yakkuma* ceremony to remove the malign attentions of *Kalu Kumara*, the demon who destroys children in the womb). Where these also fail there have always been a range of flexible kinship practices such as intra-family adoption and fostering, which have ensured that social reproduction was not confounded by the mere fact of physical infertility. What is different today is the way that these private responses to infertility become elided with broader issues of cultural survival and reproduction on the other.[5] It is in the midst of these concerns that attention begins to turn to the plethora of techniques which might 'assist' those facing involuntary childlessness.

The New Technologies: From Reproductive Crisis to Opportunity

In more recent times a new range of possibilities for combating infertility have become available in the form of the new reproductive technologies. The first IVF child was born on Sri Lankan soil in November 1999 to a Tamil couple from Batticaloa. The Sunday Times of Sri Lanka reported: 'For this history-making baby boy has not only brought hope to thousands of childless couples, but also the revolutionary "test-tube" method to Sri Lanka. And for many, the yearning and longing would be over as also the stigma of being "barren"' (14 November 1999). Whereas previously couples seeking infertility treatment would have had to travel to India, Singapore or Europe, the provision of services locally would make them cheaper and therefore more widely available to Sri Lankans. There are now two reproductive health clinics successfully offering IVF services in Colombo and a third is planned for the provincial town of Kandy. Both the Colombo clinics currently have lengthy waiting lists and have aroused a considerable amount of interest and excitement locally. Newspapers, television, mass education and regular gossip ensure that the new technologies and awareness of their consequences spread far and wide (cf. Bharadwaj 2000, 2002). In short, far more people than will ever come into direct contact with the new technologies will be left to cogitate on the wonders of modern science and what it all means for their more prosaic constructions of parenthood and personhood.

In discussion with a leading gynaecologist I asked whether he foresaw any difficulties when it came to the introduction of assisted reproduction techniques such as artificial insemination by donor. His concern with this and other techniques was not over whether there would be difficulties of introduction but how to deal with the demands made by patients for novel solutions to their reproductive problems. The apparently 'obvious' solution of drafting in brothers or other kin for purposes of donating sperm or eggs, for example, caused a western trained physician to contemplate arrangements which, in ethical terms, he found problematic. Evidently, the problem was not how to get acceptance of the solutions made possible by the new technologies but to explain why it was that they could not be used to their full extent when they provided seemingly simple solutions. As several doctors pointed out, infertility is a desperate condition and people will go to any lengths to 'get a baby'.

Among doctors and physicians, various reasons tended to be put forward for what was seen by some as a blind and worrying desire for new technologies on the part of the general public. Many identified a serious enchantment with things western in general and medical technologies and treatments in particular. Doctors as the purveyors of these new techniques occupy an exalted position. The miraculous drugs, technologies and techniques that doctors now appear to have at their disposal inspire an inordinate and, as far as some doctors are concerned, an inappropriate faith among their patients. However, the excitement and expectations which the new technologies generate among the public are often seen as unrealistic either on clinical or, as is more often the case, straightforward economic grounds. Doctors see and daily experience a significant and distressing gap between the actual and the possible.

Powerful though the dazzle of western biotechnology undoubtedly is when it comes to the new reproductive technologies, it is not the sole explanation for the positive reception of the new reproductive technologies. Technological enchantment and 'miracle' cures delivered in a largely deregulated private sector is only part of the story. To understand why the new reproductive technologies have been so enthusiastically welcomed we need also to look at the cultural spaces that have opened up and the ways in which the new technologies have slid effortlessly and logically into them. For many, the new reproductive technologies are to be welcomed because they appear to be unproblematically pro-natalist. For Buddhists in particular, the new technologies not only offer solutions to the deeply felt personal tragedy of infertility, but also provide additional opportunities for rebirth (Keown 1995: 135). At a time when concerns about religious and cultural survival are being voiced by many the possibility of reproduction, particularly when it is against the odds, begins to take on a powerful significance. Just as the horror of infertility is that there will be no sons and grandsons, no daughters and grand-daughters and, therefore, no chain of generational transmission (*paramparāva*), the appeal of the new reproductive technologies is that such transmission might be possible. However, that future, if it is to be meaningful, must necessarily be rooted in the past.

Localising a Brave New World

In trying to understand the complex relational issues that are thrown up by the new reproductive technologies, the doctors, clinicians and laypeople with whom I spoke were keen to highlight the role that local understandings of kinship and reproduction played. In the construction of plausible accounts of what, or perhaps more accurately, how, the new reproductive technologies might mean in contemporary Sri Lanka, they highlighted the way that responses to novelty and innovation might be located within a distinctively Sinhalese kinship logic.[6] The inference was that somewhere behind the enthusiastic embrace of things western lies an undercurrent of beliefs and values in relation to kinship that is closer to what Sinhalese people are really all about and this is manifest in their responses to the new reproductive technologies. In making these services available, clinicians and those charged with responsibility for overseeing the ethical regulation of new developments find themselves caught between the pull towards the uniformity of medical systems and the ethics that go with this on the one hand, and the distinctiveness of local traditions of kinship and connection on the other. Two themes that regularly featured in this regard were the practices of adoption and polyandry. Both of these are significant features of Sinhalese kinship as practiced in earlier times, they feature widely in popular traditions of how kinship used to be, as well as being formally documented within the precolonial Kandyan law (Hayley 1923).[7] Furthermore, both provide perspectives that are not readily available to Euro-American attempts to understand the new technologies.

Adoption

A common feature of Sinhalese kinship in general and Kandyan kinship in particular has been its flexibility. In his classic account of kinship and land tenure in a village in Ceylon, Leach (1961a) provided a radically materialist account of kinship, reducing it to the status of an 'idiom' for property relations – from generation to generation people move through land, not vice versa (Leach 1966). Interest in continuity of name, household and estate is paramount and groups of kin holding land or property in common will go to considerable lengths to ensure that continuity through legitimate offspring is maintained. Where legitimate offspring could not be secured by birth, the transfer of children between families appears to have been an accepted and relatively simple affair. In the past such transfers were made fully public and although biological parentage was likely to be known, little stigma appears to have attached to the child or its adoptive parents. In Kandyan Law it was stipulated that adoptions should be intra-caste – but, beyond that, little more than a public declaration or ritual statement (for example, sponsoring a puberty ceremony) was needed to effect the transfer of a child between households. The emphasis in the adoption of children appears to have been not so much on establishing legal connection as engaging in the practical business of providing care and resources; in many instances the existence of the former would be inferred from the fact of the latter. Even today, stories of children moving between families are very common.[8] However, it would seem that in recent times attitudes have begun to change with the practice having become less acceptable. For example, the law regarding public accessibility of adoption records changed in 1977 reflecting the desire for greater opacity in the movement of children between families (Goonasekere 1998: 296).

In the course of discussions about how transactions involving eggs, sperm and embryos were likely to be received in Sri Lanka, the idea that what we were dealing with was rather like adoption surfaced on numerous occasions. Passing gametes and embryos between persons was adoption of a kind, albeit at a very early stage. Thus, in discussions about Artificial Insemination by Donor (AID) the general view expressed by the group of doctors and clinicians I spoke with was that this procedure was not problematic in ethical terms at all (except for one Christian respondent for whom the introduction of semen, other than the husband's, into a woman was technically an adulterous act). Their view was that AID was probably already practiced more widely than people thought, providing a simple solution to male infertility. But, like adoption, AID could also bring potential problems. These included the management of information about genetic origins and the problem of accidental or unintended disclosure of information where parents chose to keep these secret. There were also concerns about problematic family dynamics in later life. There seemed to be some discrepancy between the views of doctors and their public in that several felt it important that sperm should originate from an anonymous donor via a sperm bank. Crucially, clinicians would be the ones who mediate these relationships with powers of selection and control lying in their hands. This arrangement would avoid resort to intra-family donation of sperm, for example, from husband's brother to brother's wife. Such transactions, although believed to be common, were seen as ethically dubious because the family would know who the biological father was and this could impact on the family in general and the child in particular in later years. However, sperm banks as such do not yet exist in Sri Lanka and the systems for collecting, screening and monitoring sperm and, in turn, matching donors and recipients are at a very early stage of development. Significantly, however, one respondent took the view that the popularly expressed position regarding intra-family donation was, in fact, the appropriate way to approach AID. Intra-familial donation would be the ideal arrangement because it was most likely to minimise jealousy and anxiety on the part of the husband. There thus appeared to be two views in circulation. In the first, the transfer of gametes and embryos, particularly if intra-family, is seen as consistent with traditional adoption practices and has a kind of continuity with traditional kinship practice. The second view, and the dominant view of the ethics committees, sees such transfers as primarily technico-medical procedures which ought to be located within the realm of professionally mediated decision making by patients. This view therefore represents a source of discontinuity with traditional ideas about the transfer of children between families with ethical regulation seemingly interposed between people's desire to achieve social reproduction on the one hand and their access to services on the other (cf. Bharadwaj 2003).

Polyandry

Although polygamy was outlawed by the British government in 1859, the practice appears to have continued, albeit with lesser frequency for another hundred years in some parts of the Kandyan Highlands (Prince Peter 1963, Tambiah 1966, Yalman 1967). Explanations put forward by anthropologists as to why this practice came into being range from sociobiological notions of parental investment through to the feudal nature of Kandyan society with its expectation that men would spend long periods in service away from home (see Hiatt 1981). However, in many of the explanations the issue of fraternal solidarity provides a common thread. Thus, for example, Tambiah (1966: 316) suggested that polyandry was a favoured option because brothers were faced with an estate that was not sufficient to support two separate wives and households. Conversely, Leach (1961b) saw polyandry solving the problem of what to do with a substantial estate which might create jealously and hostility were there to be two competing claims arising from two sets of wives and children rather than one. Either way, polyandry is associated with an expression of solidarity among men in general and brothers in particular. Suffice it to say, polyandry is nowadays extinct as a practice but it was cited by some in discussion as an explanation as to why some Sinhalese, and particularly brothers, where one of them was infertile, were so keen to offer sperm to one another. One fertility specialist said that he regularly received requests from couples to use the husband's brother's sperm to achieve pregnancy. As far as the specialist was concerned, such requests originated in a kind of kinship atavism which was clearly to be discouraged – his view of the technologies was clearly premised on heterosexual monogamy with the clinician occupying a position of considerable power in the process. But, for others, the link between intra-familial donation of sperm and polyandry as a vestige of an earlier, and perhaps more authentic kinship, was precisely the reason why allowing sperm to pass between known persons, particularly if they were brothers, was a culturally appropriate way to approach the new technologies. This is not to say, of course, that the new technologies might herald a resurgence of polyandry, but rather that the existence of this tradition provides a model for making sense of a transaction involving semen which is not available within western traditions of marriage and descent.

Conclusion

In this essay, I have offered a preliminary account of some of the culturally-specific ways in which new reproductive technologies make their appearance in contemporary Sri Lanka. I have suggested that there are links between the deep uncertainty over the country's economic and political future on the one hand, and concerns over long term demographic trends, in particular the decline in fertility rates, on the other. Concerns about the family, abortion rates and male infertility all point to deeper anxieties about the reproduction of culture, tradition and identity. Against this backdrop, the new reproductive and genetic technologies have begun to make their entrance. Indeed, these technologies and the expectations that go with them, do not simply burst unbidden onto the stage of national consciousness but are drawn along distinct institutional conduits and cultural pathways and come to occupy particular historical and political spaces. In short, the new technologies are given an overwhelming endorsement in contemporary Sri Lankan society, not simply because they address the personal tragedy of involuntary childlessness, but because they carry a collective and much weightier symbolic load. The embrace of these technologies makes the possibility of a future, quite literally, conceivable, and brings meaning and optimism at a time when the very idea of a future is clouded with uncertainty. However, as was apparent from the accounts of doctors and other members of the ethics committees, the embrace is not without its ambiguities. Although most of the doctors with whom I spoke saw the new technologies as inherently 'good', it was evident from the concerns and conflicts among them that these developments posed serious social and ethical challenges. Questions emerge about how choices and boundaries ought to be constructed within contemporary Sri Lankan society. Discussions on these matters revealed tensions between the global, hegemonising tendencies of Western medicine and the ethical systems that come as part and parcel of this process on the one hand, and the pull of local traditions which carry their own readings of personhood and the morality of connection, on the other.

Notes

1. Notable anthropological contributions in this regard have examined the discourses surrounding in vitro fertilisation (IVF) and infertility (Franklin 1995, 1997, Franklin and Ragone 1998), ideas of nature and kinship in relation to the new technologies (Strathern 1992, 1997), the analysis of prenatal diagnostic counselling (Rapp 2000), popular understandings of the new technologies (Edwards 1993, 2000), the medicalisation of genetics (Finkler 2000) and attitudes towards egg sharing and donation (Konrad 1998).
2. The research was funded by the Nuffield Foundation and carried out in Colombo, Sri Lanka in summer of 2000. (Simpson 2000, Dissanayake et al. 2002).
3. Since family planning was introduced as a national policy in 1965 there has been an increasing take up of modern contraceptive methods. In the progressive decline of the Total Fertility Rate over the last fifty years [5.3 children per woman in 1953 to 2.6 in 1985–87] contraception is estimated to have played a major role [De Silva 1996].
4. Personal communication with Sunila Abeyasekera, Director of INFORM, a leading human rights organisation in Sri Lanka.
5. Anxiety about cultural reproduction, or more accurately transmission, is amplified by the view put forward in Buddhist scriptures that the world goes through entropic cycles (*kalpaya*) in which wisdom, virtue and

morality progressively decay. Contemporary catastrophe and decline are seen by some Buddhists as entirely consistent with this model (cf. Weeratunga 2000).
6. Sinhalese kinship is a variant of the type known as 'Dravidian' (Trautmann 1981). It is characterised by a kinship terminology which classifies relatives into affines (typically those from the mother's side) and those from the father's side with whom one may not marry and who in Sinhalese kinship are referred to as 'blood kin' (*lēny āti*). Even though there is a strong patrilineal ideal or skew to the system it is essentially bilateral with children believed to acquire blood (*lē*) as well as physical characteristics from both the father's and the mother's side. Bilateral affiliation is also recognised when it comes to the transmission of property. Whilst there are some broadly unifying themes to the ideology of Sinhalese kinship, practice is subject to significant variation. Describing and explaining the variability in the uses of terminology, patterns of group formation, adherence to preferential cross cousin marriage and inheritance rules provided a major focus for early ethnographies of Sinhalese village life (Leach 1961a, Tambiah 1965, Yalman 1967, Obeyesekere 1967, Robinson 1968, Stirrat 1977). It could be argued that increased social and geographical mobility has produced an even greater diversity of kinship practices in the present day, manifest, for example, in 'love' rather than arranged marriages, neolocal residence after marriage, increased nucleation of households and the demise of large networks of kin tied together by obligations and dependencies of various kinds. However, it is still the case that traditional kinship practices occupy a significant place in people's consciousness even though they might not actually be practicing them or ever likely to practice them. They occur as one facet of a much wider nostalgia for village life evident among many Sinhalese (cf. Spencer 1991).
7. Kandyan Law is the indigenous system of customary laws that existed in the central regions of the Island until this region came under British colonial rule in 1815. Separate courts dealing with customary law were abolished shortly after this but elements of customary law were incorporated into a uniform body of law. This was particularly so in connection with laws relating to children and the family (Goonasekere 1998: 53).
8. Indeed in recent years the ease and acceptance of this practice has created significant problems in relation to transnational adoption and child abuse. Stories of families raising children specifically for adoption and 'baby farms' in isolated areas were in wide circulation. Evidence of such practices has prompted scrutiny of adoption policies and new laws to eliminate the commercial aspects of adoption (Goonasekere 1998: 135).

References

Abeyasekera, S. 1997. 'Abortion in Sri Lanka in the Context of Women's Rights'. *Reproductive Health Matters*, 9: 87–93

Bharadwaj, A. 2000. 'How some Indian baby-makers are made: Media narratives and assisted conception in India'. *Anthropology and Medicine*, 7: 1, 63–78

—— 2002. 'Conception Politics: Medical Egos, Media Spotlights, and the Contest Over the Test-Tube Firsts in India', in M. Inhorn and F. Van Balen, eds. *Infertility Around the Globe: New Thinking on Childlessness, Gender and Reproductive Technologies*. Berkeley: University of California Press, 315–33

—— 2003. 'Why Adoption is not an Option in India: The Visibility of Infertility, the Secrecy of Donor Insemination and Other Cultural Complexities'. *Social Science and Medicine*, 56: 1867–80

De Silva, W. I. 1996. 'Reproductive Change in Sri Lanka: Analysis of Intermediate Variables 1982–87', in *Social Biology*, 43: 3–4, 242–56

De Soysa, P. 2000. 'Women and Health', in S. Jayaweera, ed. *Post-Beijing Reflections: Women in Sri Lanka 1995–2000*. Colombo: Centre for Women's Research, 40–59

Dissanyake, V., Simpson, B. and Jayasekera, R. 2002. 'A Study of Attitudes Towards the New Genetic and Assisted Reproductive Technologies in Sri Lanka: A Preliminary Report'. *New Genetics and Society*, 21: 1, 65–74

Edwards, J. 1993. 'Explicit Connections: Ethnographic Enquiry in North-West England', in J. Edwards, S. Franklin, E. Hirsch, F. Price and M. Strathern, eds. *Technologies of Procreation: Kinship in the Age of Assisted Conception*. Manchester: Manchester University Press, 42–66

—— 2002. *Born and Bred: Idioms of Kinship and New Reproductive Technologies in England*. Cambridge: Cambridge University Press

Finkler, K. 2000. *Experiencing the New Genetics: Family and Kinship at the Medical Frontier*. Philadelphia: University of Pennsylvania Press

Franklin, S. 1995. 'Postmodern Procreation: A Cultural Account of Assisted Reproduction', in F. Ginsberg and R. Rapp, eds. *Conceiving the New World Order: The Global Politics of Reproduction*. Berkeley: University of California Press, 323–46

—— 1997. *Embodied Progress: A Cultural Account of Assisted Conception*. London: Routledge

—— and Ragoné, H., eds. 1998. *Reproducing Reproduction: Kinship, Power and Technological Innovation*. Philadelphia: University of Pennsylvania Press

Ginsberg, F. and Rapp. R., eds. 1955. *Conceiving the New World Order: The Global Politics of Reproduction*. Berkeley: University of California Press

Goonasekere, S. 1998. *Children Law and Justice: A South Asian Perspective*. New Delhi: Sage

Harvey, P. 2000. *Introduction to Buddhist Ethics: Foundations, Values and Issues*. Cambridge: Cambridge University Press

Hayley, F. A. 1923 (1993). *The Laws and the Customs of the Sinhalese or Kandyan Law*. New Delhi: Navrang

Heng, G. and Devan, J. 1992. 'State Fatherhood: The Politics of Nationalism, Sexuality and Race in Singapore', in A. Parker, M. Russo and P. Yaeger, eds. *Nationalisms and Sexualities*. New York: Routledge, 46–59

Hettige, S. and Mayer, M., eds. 2000. *Sri Lanka at Crossroads: Dilemmas and Prospects after Fifty Years of Independence*. Delhi: Macmillan

Hiatt, L. R. 1981. 'Polyandry in Sri Lanka: A Test-Case for Parental Investment Theory'. *Man (NS)*, 15: 582–602

Juergensmeyer, M. 1990. 'What the Bhikku Said: Reflections on the Rise of Militant Religious Nationalism (in Sri Lanka)'. *Religion*, 20: 53–75

Keown, D. 1995 *Buddhism and Bioethics*. Cambridge: Cambridge University Press

Konrad, M. 1998. 'Ova Donation and Symbols of Substance: Some Variations of the Theory of Sex, Gender and the Partible Person'. *Journal of the Royal Anthropological Institute*, 4: 4, 643–68

—— 1999. 'Technogenesis: The beginnings of microkinship.' Unpublished *ms* of paper given at the *Kinship and Temporality* conference, Goldsmiths, December 1999, n.d.

Leach, E. R. 1961a. *Pul Eliya*. Cambridge: Cambridge University Press
—— 1961b. *Rethinking Anthropology*. London: Athlone
—— 1966. 'An Anthropologist's Reflections on a Social Survey', in D. G. Jongmans and P. C. W. Gutkind, eds. *Anthropologists in the Field*. Assen: Van Gorcum, 122–47
Lock, M., Young, A. and Cambrosio, A. 2000. *Living and Working with the New Medical Technologies: Intersections of Inquiry*. Cambridge: Cambridge University Press
Martin, E. 1991. 'The Egg and the Sperm: How Science has Constructed a Romance Based on Stereo-typical Male-Female Roles'. *Signs*, 16: 495–501.
Obeyesekere, G. 1967. *Land Tenure in Village Ceylon*. Cambridge: Cambridge University Press
—— 1976. 'The impact of Ayurvedic Ideas on the Culture and the Individual in Sri Lanka', in C. Lesley, ed. *Asian Medical Systems: A Comparative Study*. Berkeley: University of California Press, 202–26
Prince Peter of Greece and Denmark. 1963. *A Study of Polyandry*. The Hague: Mouton
Rabinow, P. 1999. *French DNA: Trouble in Purgatory*. Chicago: University of Chicago Press
Rapp, R. 2000. 'Extra Chromosomes and Blue Tulips: Medico-Familial Conversations', in M. Lock et al. eds. *Living and Working with the New Medical Technologies: Intersections of Inquiry*. Cambridge: Cambridge University Press
Riessman, C. K. 2000. '"Even if We Don't Have Children": Stigma and Infertility in South India', in C. Mattingly and L. C. Garro, eds. *Narrative and Cultural Construction of Illness and Healing*. Berkeley: University of California Press
Robinson, M. 1968. 'Some Observations of the Kandyan Sinhalese Kinship System'. *Man (NS)*, 3: 402–23
Simpson, B. 2000. 'Ethical Regulation and the New Reproductive Technologies in Sri Lanka: The Perspectives of Ethics Committee Members'. *Ceylon Medical Journal*, 46: 2, 54–57
Spencer, J. 1991. *A Sinhala Village in Times of Trouble: Politics and Change in Rural Sri Lanka*. Oxford: Oxford University Press
Stirrat, R. L. 1977. 'Dravidian and Non-Dravidian Kinship Terminologies in Sri Lanka'. *Contributions to Indian Sociology (NS)*, 11: 2, 271–93
Strathern, M. 1992. *After Nature: English Kinship in the Late Twentieth Century*. Cambridge: Cambridge University Press
—— 1997. 'The Work of Culture: An Anthropological Perspective', in A. Clarke and E. Parsons, eds. *Culture, Kinship and Genes*. Houndmills: Macmillan, 48–62
Tambiah, S. J. 1965. 'Kinship Fact and Fiction in Relation to the Kandyan Sinhalese'. *Journal of the Royal Anthropological Institute*, 95: 131–73
—— 1966. 'Polyandry in Ceylon', in C. Von Fürher Von Haimendorf, ed. *Caste and Kin in Nepal, India and Ceylon*. London: Asia Publishing House, 102–40
—— 1976. *World Conqueror and World Renouncer: A Study of Buddhism and Polity in Thailand Against a Historical Background*. Cambridge: Cambridge University Press
Trautmann, T. R. 1981. *Dravidian Kinship*. Cambridge: Cambridge University Press
Weeratunga, N. 2000. 'Nature, Harmony and the Kaliyuga'. *Current Anthropology*, 41: 2, 249–68
Yalman, N. 1967. *Under the Bo Tree*. Berkeley: University of California Press

CHAPTER 3

CONCEPTION TECHNOLOGIES, LOCAL HEALERS AND NEGOTIATIONS AROUND CHILDBEARING IN RAJASTHAN

Maya Unnithan-Kumar

Introduction

As we know from the insights of significant recent work into the subject, medical technologies do not function in a moral, cultural or political vacuum. Their use is shaped by the way power relations are organised in any given context, and in turn they shape the exercise of that power. This chapter is concerned with the changing nature of social relationships which surround the use of foetal ultrasound scans, dilatation and currettage (D&C) procedures, medical termination of pregnancies and tubectomies by poor women in Rajasthan.[1] The focus is simultaneously on techniques which assist conception, as well as techniques which control conception precisely because human concerns surrounding fertility and infertility cannot be studied separately (Inhorn 1994, Kielmann 1998, Cornwall 2001, Inhorn and Van Balen 2002). The chapter considers the ways in which women and men, local healers and experts in sterility and childbirth engage with these reproductive technologies, as well as the attitudes and use made of them by the state and medical establishment. I use Rajasthani[2] women's responses to foetal ultrasound scans to suggest that the identity of mothers in relation to the foetus is constructed differently here from those contexts which are framed by less collective involvement in decisions regarding women's bodies. I suggest that the scans primarily provide women a means to validate their commitment to their spouse and affines, and have less to do with the conferral of personhood on the foetus. This finding is in contrast to the observations emerging from studies on the cultural responses to foetal imaging in the Euro-American context. For example, Petchesky (1987), Franklin (1995), Martin (1998), Mitchell and Georges (1998), Taylor (1998), have all powerfully demonstrated the role of the ultrasound scan in emphasising foetal separateness and independence from the mother, and in the construction of the baby as a person before birth, often with opposed interests to the mother.

In this chapter I suggest that despite the increasing resort to ultrasound scans, the baby continues to be regarded as a non-person until after birth in the urban periphery of Jaipur where I did my fieldwork. As I discuss, this is due to both gender ideologies and the nature of demographic realities in a region where there are high rates of miscarriage and infant mortality[3] and there is great uncertainty associated with birth. (As a result social value is accorded to children and personhood conferred once they have crossed the dangers associated with infancy; also see Scheper-Hughes 1992 for Brazil.) At the same time, the region is characterised by gender ideologies which give men primacy in procreation wherein they are regarded as creators of children with women only contributing the womb 'vessel' to carry or nurture the baby. These notions downplay some aspects of the biological connectedness between women and their children (such as the significance of the egg in procreation) compared to others (the significance of the womb). As a result women's nurturing roles become emphasised, in pregnancy and after, while men gain 'rights' through shared substance over their children along with fewer attached social commitments. As I suggest, the cultural response to foetal ultrasound testing reflects this thinking, but also shows the possibilities that exist for a renegotiation of gender and family politics.

As a means of understanding local ideas of procreation, the chapter begins by looking at the role and division of labour amongst birthing attendants and other healers. An important insight into the effects of the increasing resort to the scan and D&C procedures is provided by a consideration of the knowledge and influence of local healers such as midwives and 'spiritual' healers.[4] Paying heed to Jordan's ideas of authoritative knowledge (how one system of thinking becomes dominant) and Ginburg and Rapp's discussion on stratified reproduction (how reproduction marginalises), I show that in the specific context of Rajasthan, the authority of midwives constructed in relation to foetal ultrasound imaging is one which allows them added status in relation to other local healers concerned with childbirth, and in turn midwives have played an important role in increasing the acceptance of this technology. Local midwives' recent engagement with the scanning technology as well as with the older D&C techniques has brought them 'closer' to a biogynaecological practice which is increasingly dependent on scanning techniques for diagnoses and intervention.

In part one, I describe the ideological and reproductive context in which women seek technological intervention. In part two, I discuss: (i) women's responses to foetal images, (ii) the related issue of maternal and foetal distinctness, and, (iii) women's predisposition to further technological intervention following a scan. In part three, I use the medical interventions around contraception, such as D&C, as a means to discuss the tension between state and familial control over women's bodies, at the same time showing how women's resort to specific fertility controlling technologies are used by them as a means to negotiate better reproductive outcomes. In the concluding discussion I relate my findings to some of the broader issues around biological and social reproduction in India. But first let us turn to an understanding of the ideologies and authority that are associated with procreation and childbirth in Rajasthan.

Foetal Ultrasound, Modern Medicine and the Authority of Local Healers

I first became interested in understanding the local practices surrounding the birth of children and vulnerabilities which accompany the onset of menstruation and the reproductive span of women's lives in 1998, during one of my extended stays in the compound of a voluntary health centre on the outskirts of Jaipur city. Members of the local community surrounding the health center belonged to poor,[5] lower caste Hindu groups and to the Nagori community of Sunni Muslims.[6]

In Jaipur city and the surrounding villages, foetal ultrasound testing is locally referred to in English as 'sonography'; a term used by women who cannot read or write or even speak in English. In Jaipur the first scanner was introduced in the public women's hospital in 1981 but routine use by doctors only commenced a decade later.[7] In current biomedical practice there is an overuse of ultrasound technology with a scan being a prerequisite for any internal, especially reproductive consultations (very much along the lines described by Mitchell and Georges 1998, Taylor 1998 for America).[8] The cost of foetal scans ranged from anything between Rs 75/- (roughly equivalent to 1 pound sterling) at the main public hospital to around Rs 150/- to 350/- in the private clinics.[9] The popularity of the scanning technology amongst women relates to the increasing reliance on these methods by professional gynaecologists, the influence of local midwives, as well as the fact that such technologies promote a sense of worth associated with the value accorded by family and kin to women who bear children.[10] In this section I am specifically concerned with understanding the changing practices and attitudes of local midwives in the settlements around Jaipur city and in particular their positive response to foetal ultrasound techniques. I also discuss the ways in which there is familial based resistance to modern medicine which is epitomised in the authoritative role of spiritual healers.

I suggest that in their positive engagement with the new technologies which assist conception, local midwives neither completely negate past techniques nor completely accept current biomedical practices. They are best able to bridge the 'old' and 'new' ways of conceiving, birthing and spacing children.[11] In Rajasthan, I suggest that local midwives selectively align their skills alongside certain modern techniques which assist childbearing. Midwives are not the only local practitioners involved in facilitating pregnancy and childbirth. Whereas midwives are important facilitators at the time of birth, 'spiritual' healers (which includes a range of local healers referred to locally in terms of their practices such as *jhar phoonk*, literally sweep and blow; and *jadu-tota*, or magic and spells) are given more prominence in matters of conception, especially in so far as they deal with long term or unexplained reproductive disorders and with people's related reproductive desires and anxieties. The midwives (referred to by different local terms as discussed below), were part of a local group of women considered knowledgeable in matters of childbirth, antenatal and postnatal care, while spiritual healers, on the other hand, were distinctly associated with problems of conception, sterility and miscarriage. Local healers, including midwives, in the area where I worked believed that the younger generation had different consumption patterns and different desires, attitudes and approaches to healing compared with the older generation.

Midwives and spiritual healers in Rajasthan who were involved in facilitating childbearing were differentiated in their responses to the reproductive technologies according to their role and status in relation to childbirth. Spiritual healers were less willing to acknowledge the efficacy of biomedical techniques in general as compared to midwives. This I suggest is because of their own pre-eminence in the community as healers having an exclusive knowledge of procreation, especially in their ability to treat cases of infertility which were considered to elude biomedical practitioners. Spiritual healers were thus seen as powerful custodians of a community's reproductive knowledge (and also, as became evident to me, of their patriarchal beliefs). As elsewhere in India and South Asia (Jeffrey and Jeffrey 1993, Pigg 1997, Rozario 1998, Rozario and Samuel 2002, for example), midwives had a lower status in their communities, especially when compared to other local healers, because their knowledge was regarded as more practical, commonplace and associated with the polluting tasks of childbirth. However, in contrast to this description, in Jaipur, there were some categories of midwives who were considered authoritative in the community (in the sense used by Jordan [1997], Davis Floyd and Sargent [1997]).[12] This was especially the case for Muslim women who lived close to their natal kin thereby enabling them an access to midwives and other local healers who were natal kinswomen and men.[13] Nagori Sunni Muslim women made use of three main categories of midwives (whom I call 'kin', non-kin and city/nurse midwives) who were differentiated according to their social proximity and intimacy with the birthing mother as well as in their orientation to specialist birthing techniques and biomedicine (as I discuss in detail elsewhere; Unnithan-Kumar 2001, 2002). Considering the different ways in which midwives are categorised in this part of Rajasthan, we find that the authority constructed in relation to birthing gives precedence to an expertise coupled with social intimacy (so 'kin midwives' are more authoritative than 'non-kin midwives' who are more associated with 'pollution') and is not necessarily one that opposes traditional and biomedical techniques of birthing. The different local categories of midwives provides a good illustration of the ways reproduction is 'stratified' (in the sense used by which Ginsburg and Rapp 1995).[14]

Local midwives have always had some degree of connection with biomedical expertise, although individually they varied in their resort to such experts, depending on their assessment of their own capabilities as well as the desires of their patients.[15] Samina, Hoora and Zarina, the best regarded Sunni Muslim kin midwives I knew had all over the years developed special relationships with local private practitioners and state-trained health personnel whom they could turn to for assistance in emergencies. On their own, birthing women and their families would rarely consult or inform the state trained Auxilliary Nurse Midwife (ANM), for instance, unless they already had a long lasting acquaintance with them. Gynaecological assistance was usually sought only if the kin midwife recommended this. In 1998, I found a tendency amongst both kin and non-kin midwives to suggest to women that they undertake a foetal scan. The midwives also encouraged women to go in for D&C procedures (locally referred to as *safai*, literally meaning 'cleaning') especially after childbirth and if they were considering having another child. This was associated with the commonly held belief that the D&C would lead to the conception of a healthy foetus. Conception according to both my Sunni and caste respondents took place when the seed 'fed' off the uterus/womb (referred to as *bachhadani*, literally baby carrier). The prevailing idea was that D&C procedures helped to remove all the 'dirt' (*gandagi*) of menstruation, which was regarded as the blood of failed childbirth. D&C procedures 'cleaned' (*safai*) the uterus to enable the men's seed (*admi ka beej*), to grow in a clean (*saaf*), or unpolluted environment. An unhealthy foetus, in turn, was one that was characterised by its inability to get fat or be wet, and as 'drying out' (*sookhna*) resulting in a condition called *chhod*. The drying of the foetus was considered by kin, non-kin and city midwives alike, to result in involuntary miscarriages (*girna*, literally, falling).

Ultrasound scans provided an important means of ascertaining that the foetus was not 'drying up'. Hence it is not surprising to find that they were regularly sought out. The scans have, in turn, become a means for all categories of local midwives to substantiate their interpretations and, much like biogynaecologists, the scan has entered the routine of everyday reproductive life. Kin and non-kin midwives alike claim that the scans reinforce the outcomes of their divining techniques (*peth soothna*; literally, knowing the stomach). While the resort to foetal scanning has not changed the hierarchy operating between local midwives, it has brought them as a whole closer to the biomedical domain.

The association with biomedical techniques has also shifted the power balance between midwives and other local healers, in favour of midwives. In general though, spiritual healers continue to be regarded as authoritative with regard to the knowledge of conception.

Spiritual healers treated sterility, miscarriage and menstrual disorders and were, in contrast to local midwives, generally averse to biomedical intervention, although they too tended to see the presence of biomedicine as a sign of 'modern' times. Damodar was one such healer who belonged to the potter community. He was known in the nearby villages and in Jaipur city for his cure for sterility (a condition referred to as *banjhpan*). Damodar told me he had cured over fifty women of their inability to conceive or carry children to term. He was able to effect cures with the aid of Muslim and Hindu spirits (divine or dead beings) who possessed him. According to Damodar, it was (the being of) Ramdeo Baba who did the 'work' (*kam*) of curing sterility. In the sessions that I attended Damodar never once referred his women clients to a doctor because he said he could 'open the mouth of the womb' (*khoonk kholna*) through propitiation. Damodar did not negate the role of biomedicine, justifying biomedical intervention as an increasing necessity for those whose growth is enhanced through the consumption of modern and dangerous (*khatarnaak*) substances. According to Damodar, 'In the wheat of today, (insecticide) poison *(zehar/keede ki dawai)* is put. To kill this poison scientists made injections (*teeke*). You must use injections, because only one poison can cut (kill) another poison'. In other words Damodar believed that to counter the new forms of illness caused by the ingestion of 'new' foods, for example, wheat grown through poisonous means, one had to resort to the injections devised by scientists, as only they had the knowledge and the means to destroy the poison of their own creation.[16] However, sterility, as Damodar explained to me, was of a different making, and was a condition that was outside the control of modern medicine. Modern medicine (*angrezi dawai*; literally English medicine) was seen to be effective by healers and people alike mainly in relation to conditions such as coughs and colds, which did not have any underlying connection to spiritual agency.

Foetal scanning was locally acceptable because it reaffirmed spiritual agency at the same time as being 'non-invasive'. Vimla, a low caste Hindu woman, for example, believed that her inability to conceive after her first childbirth was connected to the dissatisfied spirit of her husband's dead male relative, who had 'tied up' the passage to her womb (Unnithan-Kumar 2003b). Although Vimla had sought a number of biomedical and spiritual cures and finally became pregnant almost eleven years after her first childbirth, she attributed her cure to supernatural rather than biomedical intervention.

There are other reasons which lead women like Vimla to reject or reluctantly accept the efficacy of biomedicine. This has to do with the attitudes which accompany the provision of biomedical care. Caste and class differences between health professionals and patients are reinforced by the new technologies and serve to emphasise the marginal status of the poor. The unsympathetic and often discriminatory treatment of poor women's reproductive disorders by medical practitioners leads women to fear and reject such encounters. Zarina, a local Sunni midwife, recounted how often her patients force her to continue with a risky childbirth because they fear discrimination from public health personnel, and are often too poor to pay high sums for immediate attention from the doctors.

Another reason for being anxious about the resort to biomedical intervention is the level of physical pain that is believed to accompany them and the invasive nature of medical diagnosis and treatment more generally (for example, through the insertion of instruments such as the speculum inside the body), which when compared to the brushing, blowing and gentle touches of the spiritual healer stands out as a physical and social violation of women's bodies. Poor women's experiences of such violation in biomedical contexts further reinforces their tendency to seek out spiritual intervention and to desire for it to be effective. Biomedical intervention is thus mainly sought when ensuring a healthy conception (through a D&C) and when affirming conception through a foetal scan.

The extent to which the advice of healers such as Damodar, or indeed of midwives, to undergo scans and D&C procedures is accepted, is by no means clear. Such advice does however play a significant role in explaining the ambivalence with regard to technological intervention.

Maternal Ambivalence and Technological Intervention

The foetal scan provided, above all, an early reassurance to Sunni and lower caste Hindu women that they were fertile and with child. (It is perhaps worth noting here that unlike in the case of women who had nutritious diets and less physically arduous lifestyles, pregnancies of poorer women were less physically visible until much later in the foetal cycle.) The resort to the scan was seen as an end in itself in terms of the proof it provided of the existence of the baby and the fertility of the mother. This is somewhat in contrast to the American women in Chicago whom Taylor (1998) interviewed, who experienced reassurance but also bonding with their babies on seeing the scan. It is also in contrast to the findings of Browner and Press (1997) who suggest that the increasing relevance of the prenatal scan predisposed the American women in their study to seek further technological intervention.

Taylor (1998) is concerned to explain the excessive use of ultrasound techniques by medical practitioners, and why, for example, it is used even for unproblematic pregnancies.[17] According to Taylor, the foetal ultrasound test has also contributed to the debates that have raged around maternal bonding (reflected in the discussions, for example, of when exactly bonding takes place). Such concerns, according to Taylor reflect, above all, societal fears regarding the emotional disposition and maternal feeling of the pregnant woman (a concern less publicly acknowledged in Rajasthan). While the discussions around maternal bonding existed before the advent of ultrasound techniques, and concentrated on the role of the birth experience as a factor in the bonding between mother and child after birth, the availability of ultrasound technology has shifted this debate to the prevalence of bonding before birth, and is used especially in the American context to persuade women planning to abort, not to do so (see Petchesky 1987, for example). As Taylor (1998) suggests, the availability of visual monitoring procedures has shifted the emphasis in the bonding debate from when, to how maternal bonding takes place. It is the promotion of maternal bonding through spectatorship which is promoted through the ultrasound techniques (Petschesky 1987, Franklin 1995, Georges 1997, Taylor 1998) rather than through the physiological and social means of bonding at childbirth. It is this shift that leads feminist social theorists such as Petchesky(1987), Martin (1987, 1998) and Rapp (1998), to suggest that medical technology and the medical profession appropriate (in their role as significant mediators) the experience of 'giving birth' to the baby. Also, the role of ultrasound techniques is instrumental in developing the bond between mother and child, which suggests that such bonds are not 'natural' (i.e., universal, biologically driven), further reinforcing the need for the technological interventions. Taylor (1998) suggests that there is a 'prenatal paradox' associated with the use of ultrasound scans – that on the one hand they help construct the tentative nature of pregnancy and the idea of the foetus 'as a commodity', at the same time as encouraging the development of a relationship between mother and baby, and an image of the foetus as a person.[18]

What my own observations suggest is that this kind of bonding does not take place in the Rajasthani context. To understand why, we have to go back to look at the specific ways in which ideas of procreation become linked to patriarchal ideologies here, where children are believed to be created from the seed of their fathers, while women only provide the womb vessel. I would suggest that the bonding that follows ultrasound scans in the American case as described by Taylor and others is absent in the case of Rajasthani women, primarily because of the belief that the baby is not 'theirs' (individually owned). That the scan mainly fulfils the role of reassurance and in fact serves to strengthen the connection between pregnant women and their affinal families.

Class and caste dimensions of doctor–patient relations embedded in the ultrasound experience also become significant in shaping women's responses to the technology and to her baby. A contributing factor to the emotional distance between a pregnant mother and her baby at the time of the ultrasound scan is the fact that, especially poorer women, are rarely invited to participate in the process by medical professionals (who consider them as 'uneducated') who operate these technologies. Vimla's experience of the scan illustrated this point. Vimla was a low caste Hindu woman in her early thirties, a time already considered locally to be too old for childbearing. Vimla was anxious as she had become pregnant after a gap of eleven years, when she had had twin daughters. She underwent a scan in the eighth month of her pregnancy and recounts:

> There was a man who operated that machine. It was turned towards him. He asked me to lie down and put some glue here (she points to the space

between her blouse and skirt). I couldn't see clearly what he was looking at. When I turned my head I could just make out some numbers and three dots and there was nothing else on the screen. No, nothing else. Obviously I could not understand what was going on in there because I can't read. Anyway it took only five minutes. After this was done the doctor asked me to get up and then placed his hand on my head and said 'Beti (daughter), what do you want?' I said a daughter or son, anything was fine. But as I already had two girls I told him I would prefer a boy. The doctor then said 'Go, your wishes will be fulfilled'. By this I guessed I would have a boy.

Vimla's account tells us how the scan reinforces her inferior social status (see Georges 1997 for similar practices in Greece), at the same time as it gives her the assurance of a child and even better, the son she and her husband are hoping for. The account also demonstrates how the sonographer acts like God, bestowing a child on her and fulfilling her desires. Although it is illegal to disclose the sex of a child in India, the sonographer is able to hint at this leading me to believe that amniocentesis is practised in some cases although data on this is difficult to gather. (It is also worth mentioning here that a boy is not necessarily always preferred over girls and that this is very much dependent on the number and sex of a couple's previous children.) Gender ideologies are also evident as reflected in the sonographers application of the lubricant in a manner that is culturally acceptable, as well as by the fact that her husband remains outside the room during this whole procedure, maintaining the appropriate conjugal decorum in public spaces. Other women with somewhat different class backgrounds may have similar experiences. Sreeja is a 29-year-old nurse from Kerala who has also had a scan in the eighth month of her pregnancy. Although she was educated and lower middle class she was also treated in an arrogant manner by her sonographer. In her case the scanner was a woman. The monitor is turned away and no explanation is offered. At the end of the scanning process, according to Sreeja, the sonographer whom she called doctor simply said, 'Yours is done. Go. Come in the morning and take the report'. Sreeja's husband had waited outside the consultation room.

When we juxtapose Vimla's experience of the foetal ultrasound with the description of similar procedures in the United States and Greece (Mitchell and Georges 1998) we can see a significant difference emerge. According to Mitchell and Georges, an important part of the scanning routine is when the sonographers describe the image and 'show' the baby to the expectant parents. It is their talk about the baby, its physical appearance, its moods, its actions, its social connectedness to the parents, that transforms the ultrasound image into a culturally meaningful baby (1998: 108).

The absence of any information related to the baby while viewing the scan results in women remaining distanced from such proceedings and from any direct engagement with the baby. The low degree of involvement of women in interpreting the scan may prevent them from experiencing what Georges (1997) in her study of the impact of foetal ultrasound imaging in Greece suggests is the visual pleasure associated with seeing the foetal image. In the Greek case the foetal images serve to heighten the physical awareness pregnant women have of their babies and thus 'reconfigure the structure of feeling of pregnancy'. In the Indian case, the scanning technologies mainly serve to maintain the high status of the doctors who operate them and the midwives who are associated with them. Technologies of procreation thus remain awesome and forbidden for the rural Sunni and Hindu women and serve to maintain the exotic and alien nature of a medical system 'foreign' to them (the *angrezi* or 'English' aspect of them).

The responses to the ultrasound technologies are ambivalent and a complex mix of what the women engaged in these processes want for themselves, how they view their social responsibilities to their husband and his family, as well as their own previous experiences of the treatment. This ambivalence is reflected in the use of scans to verify pregnancy but with little recourse to further technological intervention (also see Shaw, this volume, for a similar response to genetic testing). In most of the cases which I came across, where women had gone in for ultrasound scans, I found that this did not predispose them to seek further biomedical intervention in childbirth. This finding is in contrast to the conclusion reached by Browner and Press (1997) in their study on American women's responses to biomedical interventions around prenatal care. What Browner and Press suggest is that, before the advent of the scanning technologies, when these interventions were not confirmed by embodied knowledge and experience, or could not be easily incorporated into their daily schedules, the American women in their study tended to reject biomedical prescription. However, with the increasing use of technological aids such as the ultrasound scans, Browner and Press found that it was becoming increasingly difficult for the women to resist biomedical intervention in prenatal care. In other words, resort to the ultrasound scans predisposed women to seek further biomedical intervention. In the case of poorer women in Rajasthan, it was not simply a question of finances (as reproductive matters may receive priority funding even in relatively poor households) but because the kind of consumption represented by the ultrasound scan was less tied to images of modernity (as among the wealthier middle classes; Donner, this volume) and more to doctors attitudes and familial control, as Miriam's account below shows.

Miriam, a Nagori Muslim, was in her mid-thirties and pregnant with her ninth child when I met her in 2000. A kin midwife, Samina, had diagnosed that the baby was in breech position and this was confirmed by a foetal scan. Miriam was known to have a difficult child labour as was her experience in all her previous pregnancies and her anxiety was heightened by the breech condition. As on previous occasions, she consulted with a range of spiritual healers from within her husband's family (her father-in-law) and from other castes (like Pooran the healer from the 'untouchable' caste). She also sought the advice of Suman, a gynaecologist in the state hospital, ten days before she gave birth because she had been experiencing pain and some fluid loss. Suman advised Miriam to collect her things from home and come back to be admitted into hospital. When Miriam got home that day her father-in-law became possessed by Shukker Baba (a locally revered saint known for his healing powers) who pronounced that he would either kill her or her baby if she went to the hospital. Baba said that the cause of her pains lay in the fact that her husband's elder brother, who had died shortly after being born, was seeking recognition by the family. Miriam decided to heed Baba's advice and remained at home. She later gave birth to a baby girl with the help of Farzan, a kin midwife and Saira, a neighbour.

Miriam's example above all demonstrates how the conjunction between healing and patriarchal, especially affinal control (reflected in her father-in-law's threats), serves to define the context in which poor women respond to biomedical intervention. Class or consumption desires become one amongst a range of other factors which determine women's resort to technological intervention. Rapp (1998) makes a similar observation about the number of crosscutting factors such as religious affiliation, kinship, class which structure women's refusal to undergo pre-natal testing in New York. In the next section I examine in greater detail women's resort to contraceptive technologies as a response to their own reproductive desires, as well as to the reproductive expectations placed upon them by their spouses, families and the state.

Safai and Women's Negotiation of Familial Control

In this section I focus on the role kinship and familial authority play in defining women's reproductive agency. Such agency may resist or coincide with family and state health agendas to do with conception. At the level of gender relationships I suggest that the resort to contraceptive technologies are both constrained by ideas of what it means to be a good wife and mother and notions of self and collective responsibility. At the same time the resort to biomedical intervention also plays a part in reconfiguring and re-imagining notions of self and gender. In order to illustrate what I mean, I will focus on two local concepts *safai* (literally cleaning, the local term for D&C techniques) and *girna* (literally falling, referring to involuntary miscarriages) and in particular on the dual meanings attributed by local women to the former.

The popular demand for D&C techniques shows that these procedures were less hedged with ambivalence compared to any other biomedical intervention. As mentioned previously, both kin and non-kin midwives as well as professional gynaecologists promoted the use of D&C techniques (which they all referred to as *safai*). For midwives, the D&C procedure cleaned away the 'dirt' left behind by previous cycles of 'failed childbirth' (as manifest in menstruation as well as in miscarriages) and was a means of assuring a healthy pregnancy. Professional gynaecologists also routinely recommended a D&C following a miscarriage due to the biomedical concern with reproductive tract infections. (The dangers associated with the overzealous D&C procedures were not recognised locally. For example, there were some gynaecologists who suggested [confirming my own suspicions] that a vigorous scraping of the uterus wall was in fact among the significant causes of sterility in women.)

The D&C technology when compared to the foetal scanning technology had been longer in use, at least by a decade or so, was less expensive and was more routinised in reproductive health seeking behaviour compared to the latter. There was a surprising disconnection in the use of the two technologies, i.e., I found that D&C procedures were usually resorted to in the very early stages of pregnancy, compared to the scans and did not follow on from the scans (except in the more exceptional cases of sex selective abortions).[19]

The resort to D&C procedures provided greater opportunities for resistance to the dominant ideology compared to the scan, for example, allowing women to negotiate their own reproductive terms within the cultural framework available to them. Here the double meaning attached to *safai* (as referring to both D&C and aborting procedures) becomes important. During the course of my stay in Jaipur, it became clear to me that the term *safai* was not only used to refer to D&C procedures but also referred to abortion, i.e., voluntary miscarriage.

The significance of referring to *safai* in terms of the termination of pregnancies becomes clear when seen especially in relation to another local concept, that of *girna* (literally to fall, referring to involuntary miscarriage). *Girna* was a fairly frequent occurrence amongst the women I got to know. Nearly all of the fifty-five women I interviewed reported at least two miscarriages in their reproductive history.[20] But as I was to find out, the boundaries between *safai* (in its sense as an abortion or induced and voluntary miscarriage) and *girna* (or involuntary miscarriage) were blurred. I found that some women claimed having a miscarriage (*girna*) as a reason for seeking a D&C (*safai*) but in fact doctors could find themselves aborting a foetus in its early stages.[21] When I questioned women about the signs which indicated to them that they had had a miscarriage (*girna*), there was some vagueness suggesting the occurrence of a bit of blood spotting to a heavier bleeding. However, the occurrence of any bit of bleeding was enough grounds for women to claim they had had an involuntary miscarriage (*girna*) and demand the finances for a D&C (*safai*) from their affinal family.

The idea that D&C procedures ensured a healthy environment for the foetus to develop, was one of the primary reasons for its acceptance among affinal kin. The husband's family will readily provide the necessary finances to undertake procedures which they believe secure the birth of their child. On the other hand, through a resort to D&C procedures, women can both gain the social value from becoming pregnant and also not necessarily carry a child to term (also see Day for a very similar tactic among London sex workers, 2001).

Contraceptive technologies most of all threatened familial and patriarchal interests in women's reproduction. Reproductive technologies such as the D&C and the medical termination of pregnancies were used by individual women (especially those who already had several children) as a means to space their children. Women's claims of involuntary miscarriage were rarely contested because it was information that only they had knowledge of. Under the guise of seeking the *safai* as a D&C related intervention, women were able to exercise their choice with regard to the bearing and birthing of children, an important choice given the fact that birth control is collectively frowned upon and a difficult issue for women to broach with their husband and his family, unless initiated by the husbands themselves. But clearly not all women felt in a position to interpret a bleeding during pregnancy as an indication of a miscarriage. Individual women felt constrained by the obligations and emotions that tied them to their husbands, and it was usually women who had more difficult relationships with their husbands who resorted to such measures (Unnithan-Kumar 2001, 2003a). In this sense the reproductive technologies available enabled women to use their childbearing as a means of negotiating the familial and spousal expectations placed upon them. The next section considers the extent to which women's resort to technological intervention in childbirth is linked to state and medical agendas.

Public and Private Medical Services

On the face of it Sunni and caste women's recourse to D&C procedures would seem to support a state health programme dominated by a rationale for population reduction and family planning. However, in reality this was not the case as the state favoured terminal rather than temporary fertility reduction measures, as reflected in the continuing emphasis on tubectomies in government health programmes (Visaria and Ramachandran 2001). In fact here we find public doctors and health workers in collusion in inflicting tubectomies on women. A popular perception among poor Sunni and caste women of the reproductive services provided by the public hospitals in Jaipur was that if you went in for a medical termination, you would find yourself sterilised at the end of the procedures. In general, public health services were regarded as associated with contraceptive coercion. Village health workers, such as the ANM, for example, were associated mainly with the work around family planning, in distributing condoms, intra-uterine devices (IUDs) and hormonal pills, rather than with their skills in childbirthing.

Despite the coercion associated with medical treatment in the public hospitals there was a demand for terminal methods of contraception. This acceptance of biomedical intervention was limited to those women who were well advanced in their reproductive cycle and who had children of both sexes. Older Sunni and caste women who had more than five children with at least one son would all consider undergoing a tubectomy (locally referred to by the English word 'operation'). Whether they actually went in for one depended on a number of factors such as their ability to command household finances, alternative domestic support as well as the disposition of their husbands towards their condition (Unnithan-Kumar 1999, 2003a). In some cases these women were encouraged by their husbands to go in for more permanent methods of contraception, especially where several children were already present and the household was poor. As a result we also find an interesting division in health seeking behaviour, wherein poor and older women went to public hospitals to undergo tubectomies. In contrast, younger women who sought D&C procedures went to private clinics and hospitals.[22]

The resort to private clinics by younger women was especially reinforced by the association of these places with fertility enhancing techniques (much as in Italy, as Bonaccorso describes in her chapter). In the everyday context of Nagori Sunni women's lives secondary infertility (i.e., the inability to conceive after having the first child), was a much more visible concern and hence there was more of an active interest in seeking out interventions which assisted conception. Private medical institutions not only provided the services that Nagori men and women actively sought but they did so through resort to the latest technology, with due time given to each patient and in an efficient manner (for example, with tests results made available within the times stated). This is in stark contrast to the poor conditions of the public hospitals, the long delays in meeting the doctors who are largely unsympathetic and often coercive in promoting contraceptive techniques, and the hidden costs of services provided in these places. It is then not difficult to understand why even women from poorer households (having an average earning of around Rs 1000/- for a household with 6–7 members) will seek out consultations in private clinics and hospitals despite costs of Rs 100/- per consultation (and scans which cost upto Rs 300/-). Affinal families were willing to provide the finances though and here, as with the D&C procedure, we see a greater meeting of women's reproductive desires with that of their affines reinforced through the ultrasound scans.

Concluding Comments

The Significance of Class

Any experience of medicalisation in India can only be understood in the context of caste, class and gender ideologies and practices. Class, as well as race, age and sexual preference, have been shown to shape women's responses to techniques and technicians of assisted birth in the Euro-American context (Martin 1987, Fraser 1995, Rapp 1998, Ragone 2000, Becker 2001, for example). In Northern Europe and North America we find that it is middle class women who most resist biomedical interventions because they value the ability to participate in the birth of their babies and have the means to seek birthing care outside the mainstream hospital and clinics. Working class mothers may neither have the means to go in for the more expensive alternative options of childbirth nor necessarily reject biomedical intervention often preferring biomedical assistance such as pain relief for easy and quick births. (This is of course an over generalisation when seen in the context of Rapp's nuanced work, 1998.) In the Indian context this distinction in class preferences is not as stark, with both lower and middle class women appreciating biomedical intervention (for the latter see Donner, this volume). There is however a stark class distinction in the manner women from the middle, upper and lower classes are treated. Poorer women approach biomedical interventions more fearfully because of the very real dangers that its unethical practice presents them and because of their lack of command over the social agents that operate these interventions (also see Ram 1998, Van Hollen 2002). Often the poorest women are those who belong to the lowest castes, and face caste discrimination from public and private doctors alike who overwhelmingly belong to the middle and upper castes. Middle and upper-middle caste and class women are less likely to face such marginalisation and thus are more likely to approach technological intervention in terms of the practical and emotional benefits it confers them. They also have the finances to opt for costly Caesarean sections (Donner, this volume). While both wealthier and poorer women are subject to the controls of their family and religious ideologies, among poorer women, age and number of children further determines the extent to which a woman has the power to decide on her childbearing capacity and resort to technological intervention.

The Focus on Fertility

Issues relating biological to social reproduction have never really been at the theoretical core of understanding social change in India.[23] There are some recent noteworthy exceptions (for example, Patel 1994, Gold 1994, Das 1995, Jeffrey and Jeffrey 1993, 1997, Pigg 1997, Ram 1998, Rozario and Samuel 2002, Van Hollen 2002, Unnithan-Kumar 2001, 2002, 2003). The focus on reproductive change in Indian society has mainly been framed by the work on 'fertility', in particular in its role in characterising the nature of India's demographic transition. Demographic and public health understandings of fertility which give primacy to its biological meaning have both dominated the arena of population and development planning and also framed local understandings of progress in India. Although accepted at the level of community and household, these understandings are not always reflected in practice (as seen in the many instances of popular resistance to the Family Planning programmes of the state). Nevertheless, over the past decade and a half, powerful critiques of the standard demographic and development perspectives have emerged from amongst development practitioners, feminists in the various disciplines and amongst health activists (notably Greenhalgh 1995, Kertzer and Fricke 1997 and Bledsoe 2002, more generally, and for India, Dyson and Moore 1983, Gittelsohn et al. 1994, Sen et al. 1994, Patel 1994,

Dasgupta et al. 1998, Chayanika et al. 1999, Ramasubban and Jejheebhoy 2000, Visaria and Ramachandran 2001, Ravindran and Panda 2002, Kielmann 2002). These critiques stem from more locally framed understandings of reproductive change and social transformation. The recent feminist critiques of family planning interventions in Rajasthan (for example, Visaria and Visaria 2001) are significant in that they draw upon empirical data to point to some of the bottlenecks and structural gaps in the health care sector in the state. While highly pertinent to policy and planning exercises, these reports do not, however, directly engage with the nature of the ideologies and practices to do with reproduction in the region, which I would suggest is equally vital in thinking about the trajectories of reproductive change. A focus on local notions and attitudes to childbearing and the related response to certain reproductive technologies which assist conception enables us to understand, for example, that concerns of fertility cannot be understood apart from those of infertility (what Inhorn calls the fertility-infertility dialectic; 1994, Inhorn and van Balen 2002). As this chapter has discussed, the cultural importance given to fertility and childbirth, and the related stigmatisation of infertility is especially reflected in Rajasthan in the high status and position of local healers who facilitate birth and cure infertility.[24]

The Role of the State and Private Medical Practice

A major consideration relating to reproductive change in Rajasthan is related to the culture of state planning and medical practice in India and the different ways in which public and private medical practitioners in India approach, use and promote the reproductive technologies and in turn construct pregnant women, midwives and birth very differently from the way they are perceived at the local level. In general, there is a clear division between the promotion of techniques for women which facilitate conception (foetal scans, D&C and IVF treatment)[25] and those which control conception (tubectomies, condoms, intra-uterine devices, hormonal pills, injectables), with the latter promoted by the government health programmes and the former associated with local healers and private clinicians. Abortion services are provided by legally authorised public and private health centres but also illegally by private clinics and certain specialist 'city' midwives. [26] Given the local emphasis on childbearing and the social stigma related to infertility, state efforts at promoting contraceptive techniques in Rajasthan are met with local avoidance. Women who do seek out contraceptive methods are mostly women who are past their major childbearing years (Unnithan-Kumar 1999, 2001). As I discuss in the chapter, certain reproductive techniques, such as D&C, abortion and tubectomy, can be used by individual women to 'space' their babies or by older women who have 'proved their worth' and achieved the desired number and sex of their children, to terminate their childbearing capacities. Women may resist familial pressures to conceive but may also be constrained by notions of spousal and familial loyalty and obligation from exercising such resistance and choice (i.e., in deciding to space rather than terminate their childbearing potential).

The state's uptake of reproductive technologies in its health programmes in India reflects how it is caught between the attempts to regulate a private sector which provides easy access to technological interventions and, its own 'development' driven agenda to control fertility. In its recourse to technologies which promote conception, the private health sector, on the other hand, is able to address the significant local anxieties surrounding infertility. It thus provides, even poor women with access to the latest technologies of conception. However, access to the latest technologies provided by private medical institutions comes at a cost, which especially for poorer men and women is manifest with the risks associated with the fraudulent and unethical nature of private practice where there is little if any means of redress.

To summarise, the focus on the use of reproductive technologies such as ultrasound scans and D&C procedures in Rajasthan as discussed in the chapter lead me to suggest that the introduction of modern reproductive techniques can enhance the authoritative knowledge on childbirth held by both gynaecologists and local midwives alike. In their introduction to non-Euro-American contexts, new technologies which assist conception need not undermine the existing authority of indigenous healers such as kin midwives. Kin midwives are seen to advocate the use of select technologies such as ultrasound scans and therein are seen to strengthen their own standing within the communities where they practice. In terms of the responses of childbearing women to the ultrasound techniques, we find that their popularity and use among poorer women in Rajasthan has not resulted in a greater disposition of these women to take up further biomedical assistance in childbirth. Although the recourse to D&C has always been high, there is little perceptible increase in the use of D&C procedures following the popularity of the foetal scans. The foetal ultrasound scan functions as an end in itself: providing a reassurance of the child bearing capability of women. The scans do not result in an increase in maternal bonding or in the construction of the foetus as a person before birth, partly because of local ideas of conception and partly because of poor communication between scanners and pregnant women. The bonding between a woman and her affines may, however, increase as a result of a positive scan. The distinctive responses of poor Rajasthani women to the scan and the older D&C procedures are best understood when considered in terms of the high value placed on fertility, on 'healthy' conception, the anxieties of infertility, individual concerns for spacing children, the violence associated with biomedical techniques, and local notions of procreation and gender responsibilities wherein men are regarded as creating and thus 'owning' babies and the women as carrying them to term. Familial authority amongst caste and Sunni Muslim communities alike ensures a cautious predisposition to the technologies which facilitate conception and a resistance to state sponsored contraceptive technologies. Individual women and men are nevertheless able to use the reproductive technologies to differing extents to their own advantage.

Notes

1. It is important to remember that reproductive technologies are a differentiated category and accordingly the responses to them will be varied.
2. I use the word Rajasthani to denote only a general affiliation (in terms of a shared regional language, history and state government) between people who live in Rajasthan. In my use of the term, I do not imply that cultural differences between the various communities and social groupings be collapsed.
3. According to the most recent National Fertility and Health Survey of 1998–99, during the five years preceding the survey, the infant mortality rate was 80 deaths of infants between 0 to 11 months per 1000 live births.
4. The term spiritual healer is used here as a general term to represent a variety of healers who deal with sacred images, objects and supernatural beings.
5. There is obviously a vast range of people who can be categorised as poor. In my use of the term poor here, I refer to households with an average of six members and a monthly income of between Rs 1000/- to Rs. 1500/- (approximately 14 to 21 pounds sterling).
6. In general, I found very similar ideas and practices of procreation, birth and maternal care operating across the lower caste Hindu and Muslim communities. There were also some interesting differences. The differences in the approaches to childbirth and maternal care were mainly to do with the different patterns of marriage and residence, resulting in different, yet significant, roles of the family in each case (Unnithan-Kumar 1999).
7. Personal communication with gynaecologist Suman.
8. One senior gynaecologist at the central hospital in Jaipur told me that she referred nearly every single patient to the scan as this made her diagnosis more authoritative.
9. There was only one scanner in use at the public hospital where roughly 80–90 patients were referred to in a day, making it accessible to those who were willing to endure long queues, usually the very poor.
10. Yet this is a value related to carrying rather than creating children as it is men who are regarded as creating children through their sperm while women merely provide the 'vessel' to nurture the baby. Men are held as powerful in their role as procreators and at the same time also absolved of all responsibilities for a failure of that process (also see Inhorn 1994 for a very similar situation in Egypt).
11. In her study of African-American midwives, Fraser (1995) makes a similar point about the ways in which connections are made between what seem to be parallel worlds of old and new. Fraser's informants (including the midwives she interviewed) do not completely invalidate the past techniques of childbirth, but rather render them unworkable in the present context because they believe that modern bodies and minds are differently constituted. As a result, the skills of African-American midwives are dying out with the advent of modern medicine.

12. Jordan (1997: 57) describes authoritative knowledge as 'the way of organising power relations in a room that makes them seem literally unthinkable in any other way ... that the power of authoritative knowledge is not that it seems correct, but that it counts.
13. Another difference was in the Sunni Nagori practice of the *ajaan,* or conferral of a name and identity to the baby soon after birth, which is much sooner compared with rituals of conferral of social identity (such as the naming or haircutting ceremonies) among lower caste groups in the region.
14. According to Ginsburg and Rapp (1995: 3), the term 'stratified reproduction' is taken to describe power relations by which some categories of women are empowered to nurture and reproduce, while others are disempowered or marginalised from such processes.
15. Zarina, an established kin and 'city' midwife explained her frequent reluctance to seek assistance in terms of the lack of power she felt in the face of poor clients who insisted she alone complete the delivery of the child as they had no funds to turn elsewhere.
16. In Damodar's narrative there is also a distinct tendency to view illness as a modern phenomenon linked to changing lifestyles which, in his case, were characterised as increasingly less physically arduous. Damodar said he himself was less prone to such modern ailments because of the physically laborious nature of the tasks he performed, as well as the fact that he ate only what he produced. He went on to suggest that my research assistant and I were frequently ill because, like other city dwellers, we 'grazed like goats' and did not 'work', unlike the poor.
17. Ultrasound, Taylor (1998) observes, is used both as a means of prenatal diagnosis to see if the foetus is healthy but also to make decisions about the medical management of pregnancy, birth, abortion. In most cases, however, the ultrasound will not reveal foetal abnormalities and its use therefore is more a projection of desires and anxieties relating to the outcome of pregnancy than anything else.
18. The sonogram enables links to be formed between the foetus and not just with the mother but with the wider kin group. In this sense the sonogram becomes a kind of 'shared substance of kinship' along the lines in which Schneider talks about blood and substances of procreation.
19. I would like to thank Stanley Ulijaszek for alerting me to the relationship between the D&C and foetal scans, and other members of the Oxford fertility and reproduction group for their helpful comments on the subject.
20. Most women had two to three cases of involuntary miscarriages while several had upto four such experiences. The lower nutritive intake of these women and the high prevalence of accompanying anaemia would provide further agreement with these findings.
21. Here we see an example of a situation where doctors are used by women clients to meet their own reproductive needs. There is also nevertheless a willingness on the part of doctors to perform abortions based on the common belief that the fertility of the illiterate and poor needs to be controlled.
22. There was also a small but growing trend among younger women in the area to try intra-uterine devices such a copper T, hormonal pills and condoms, in that order. Such resort was constrained, however, by the 'progressive' nature of the husband as well as the social intimacy and sense of loyalty women felt towards their spouse.

23. Such understandings have been rooted elsewhere, in the studies of urbanisation and the implications this has had for the institutions of caste and joint family, for example. More recently, the role and meaning of the state, the wider economic and political processes and the role of religious conflict in particular frame the discussions of social change in India (Van der Veer 1996, Fuller and Benei 2001, Hansen and Stepputat 2001).
24. A focus on local concerns also highlights the role of marital structures and sentiments in determining the responses to the medical technologies of conception and contraception (see for example, Pearce 1995 for Nigeria).
25. In vitro fertility or IVF treatment although gaining in popularity among the middle and upper-middle class women is still inaccessible to poorer women. These services are mainly provided by private clinics and hospitals. In Jaipur in 2000 there were two such clinics which were well known. Poorer women frequented these clinics and private hospitals for consultation with the doctors and for ultrasound scans.
26. Since 1979, abortion has been legalised in India, allowing pregnancies to be medically terminated only by authorised medical personnel.

References

Becker, G. 2001. *The Elusive Embryo: How Women and Men Approach New Reproductive Technologies*. Berkeley: University of California Press

Bledsoe, C. 2002. *Contingent Lives: Fertility, Time and Aging in West Africa*. Chicago: University of Chicago Press

Browner, C. and Press, N. 1997. 'The Production of Authoritative Knowledge in American Prenatal Care', in R. Davis-Floyd and C. Sargent, eds. *Childbirth and Authoritative Knowledge*. Berkeley: University of California Press, 113–32

Chayanika, Swatija and Kamaxi. 1999. *We and Our Fertility: The Politics of Technological Intervention*. Mumbai: Comet Media Foundation

Cornwall, A. 2001. 'Looking for a Child: Coping with Infertility in Ado-Odo, South-Western Nigeria', in S. Tremayne, ed. *Managing Reproductive Life*. Oxford: Berghahn Books, 140–57

Das, V. 1995. 'National Honour and Practical Kinship: Unwanted Women and Children', in F. Ginsburg and R. Rapp, eds. *Conceiving the New World Order*. Berkeley: University of California Press, 212–34

Dasgupta, M., Chen, L. and Krishnan, T. N., eds. 1998. *Women's Health in India: Risk and Vulnerability*. Delhi: Oxford University Press

Davis-Floyd, R. and Sargent, C. 1997. *Childbirth and Authoritative Knowledge: Cross-Cultural Perspectives*. Berkeley: University of California Press

Day, S. 2001. 'Biological Symptoms of Social Unease: The Stigma of Infertility in London Sex Workers', in S. Tremayne, ed. *Managing Reproductive Life*. Oxford: Berghahn Books, 85–103

Dyson, T. and Moore, M. 1983. 'On Kinship Structure, Female Autonomy and Demographic Behaviour'. *Population and Development Review*, 9: 1, 35–60

Edwards, J. 2000. *Born and Bred: Idioms of Kinship and the New Reproductive Technologies*. Oxford: Oxford University Press

—— Franklin, S., Hirsch, E., Price, F. and Strathern, M. 1993. *Technologies of Procreation: Kinship in the Age of Assisted Conception*. London: Routledge.

Franklin, S. 1995. 'Postmodern Procreation: A Cultural Account of Assisted Reproduction', in F. Ginsburg and R. Rapp, eds. *Conceiving the New World Order.* Berkeley: University of California Press, 323–46

Fraser, G. 1995. 'Modern Bodies, Modern Minds: Midwifery and Reproductive Change in an African American Community', in F. Ginsburg and R. Rapp, eds. *Conceiving the New World Order: The Global Politics of Reproduction.* Berkeley: University of California Press, 42–59

Fuller, C. and Beneie, V. 2001. *The Everyday State and Society in Modern India.* London: Hurst and Company

Georges, E. 1997. 'Fetal Ultrasound Imaging and the Production of Authoritative Knowledge in Greece', in R. Davis-Floyd and C. Sargent, eds. *Childbirth and Authoritative Knowledge.* Berkeley: University of California Press, 91–113

Ginsburg, F. and Rapp, R. 1995. *Conceiving the New World Order: The Global Politics of Reproduction.* Berkeley: University of California Press

Gittelsohn, J., Bentley, M., Pelto, P., Nag, M., Pachauri, S., Harrison, A. and Landman, L., eds. 1994. *Listening to Women Talk about their Health: issues and Evidence from India.* Delhi: Ford Foundation

Gold, A. 1994. 'Sexuality, Fertility and Erotic Imagination in Rajasthani Women's Songs', in G. Raheja and A. Gold, eds. *Listen to the Heron's Words.* Berkeley: California Press, 30–73

Greenhalgh, S. 1995. 'Anthropology theorises reproduction: Integrating practice, political economic and feminist perspectives', in S. Greenhalgh, ed. *Situating fertility: Anthropology and demographic enquiry.* Cambridge: Cambridge University Press, 3–29

Hansen, T-B. and Stepputat, F., eds. 2001. Introduction. *States of Imagination: Ethnographic Explorations of the Postcolonial State.* Durham: Duke University Press, 1–41

Inhorn, M. 1994. *Quest for Conception: Gender, Infertility and Egyptian Medical Traditions.* Philadelphia: University of Pennsylvania Press

—— 2000. 'Missing Motherhood: Infertility, Technology and Poverty in Egyptian Women's Lives', in H. Ragone and F. Winddance Twine, eds. *Ideologies and Technologies of Motherhood.* New York: Routledge

—— and Van Balen, F., eds. 2002. *Infertility around the Globe: New Thinking on Childlessness, Gender and Reproductive Technologies.* Berkeley: University of California Press

Jeffrey, P., Jeffrey, R. and Lyon, A. 1989. *Labour Pains and Labour Power: Women and Childbearing in India.* London: Zed Books

Jeffrey, R. and Jeffrey, P. 1993. 'Traditional Birth Attendants in Rural North India: The Social Organisation of Childbearing', in S. Lindenbaum and M. Lock, eds. *Knowledge, Power and Practice: The Anthropology of Medicine and Everyday Life.* Berkeley: Univerisity of California Press, 7–32

—— and Jeffrey, P. 1997. *Population, Gender and Politics: Demographic Change in Rural India.* Cambridge: Cambridge University Press

Jordan, B. 1997. 'Authoritative Knowledge and its Construction', in R. Davis-Floyd and C. Sargent, eds. *Childbirth and Authoritative Knowledge.* Berkeley: University of California Press, 1–55

Kertzer, D. and Fricke, T. 1997. *Anthropological Demography: Toward a New Synthesis.* Chicago: University of Chicago Press

Kielmann, K. 2002. 'Theorising Health in the Context of Transition: The Dynamics of Perceived Morbidity among Women in Peri-urban Maharashtra, India', in *Medical Anthropology*, 21: 2, 157–207

Martin, E. 1987. *The Woman in the Body: A Cultural Analysis of Reproduction.* Boston: Beacon Press
—— 1998. 'The Fetus as Intruder: Mother's Bodies and Medical Metaphors', in R. Davis-Floyd and J. Dumit, eds. *Cyborg Babies: From Techno-sex to Techno-tots.* New York: Routledge, 125–42
Mitchell, L. M. and Georges, E. 1998. 'Baby's First Picture: The Cyborg Fetus of Ultrasound Imaging', in R. Davis-Floyd and J. Dumit, eds. *Cyborg Babies: From Techno-sex to Techno-tots.* New York: Routledge, 105–24
Patel, T. 1994. *Fertility Behaviour: Population and Society in Rajasthan.* Delhi: Oxford University Press
Pearce, T. O. 1995. 'Women's Reproductive Practices and Biomedicine: Cultural Conflicts and Transformation in Nigeria', in F. Ginsburg and R. Rapp, eds. *Conceiving the New World Order.* Berkeley: University of California Press, 195–209
Petchesky, R. 1987. 'Fetal Images: the Power of Visual Culture in the Politics of Reproduction', in M. Stanworth, ed. *Reproductive Technologies: Gender, Motherhood and Medicine.* London: Polity Press, 57–80
Pigg, S. 1997. 'Authority in Translation: Finding, Knowing, Naming and Training "Traditional Birth Attendants" in Nepal'. R. Davis-Floyd and C. Sargent, eds. *Childbirth and Authoritative Knowledge: Cross-Cultural Perspectives.* Berkeley: California University Press
Ragoné, H. 2000. 'Of Likeness and Difference: How Race is being Transfigured by Gestational Surrogacy', in H. Ragoné and F. Winddance Twine, eds. *Ideologies and Technologies of Motherhood,* 56–76
Ragoné, H. and Winddance Twine, F. 2000. *Ideologies and Technologies of Motherhood: Race, Class, Sexuality, Nationalism.* New York: Routledge
Raheja, G. and Gold, A. 1994. *Listen to the Heron's Words: Reimagining Gender and Kinship in India.* Berkeley: University of California Press
Ram, K. 1998. 'Maternity and the Story of Enlightenment in the Colonies: Tamil Coastal Women, South India', in M. Jolly and K. Rams, eds. *Maternities and Modernities.* Cambridge: Cambridge University Press, 114–44
Ramasubban, R. and Jejheebhoy, S., eds. 2000. *Women's Reproductive Health in India.* Jaipur: Rawat
Rapp, R. 1998. 'Refusing Prenatal Diagnosis: The Uneven Meanings of Bioscience in a Multicultural World', in R. Davis-Floyd and J. Dumit, eds. *Cyborg babies: From Techno-sex to Techno-tots.* New York: Routledge, 143–67
Ravindran, T. K. and Panda, M. 2002. 'Gender and Fertility Transition in the South Asian Context'. Paper presented at the conference on Gender and Health in South Asia. Heidelberg, July 2002
Rozario, S. 1998. 'The Dai and the Doctor: Discourses on Women's Reproductive Health in Rural Bangladesh', in K. Ram and M. Jolly, eds. *Modernities and Maternities: Colonial and Postcolonial Experirneces in Asia and the Pacific.* Cambridge: Cambridge University Press
—— and Samuel, G. 2002. *The Daughters of Hariti: Childbirth and Female Healers in South and Southeast Asia.* London: Routledge
Scheper-Hughes, N. 1992. *Death Without Weeping: The Violence of Everyday Life in Brazil.* Berkeley: University of California Press
Sen, G., Germain, A. and Chen, L., eds. 1994. *Population Policies Reconsidered: Health, Empowerment and Rights.* Boston: Harvard school of Public Health.
Schneider, D. 1984. *A Critique of the Study of Kinship.* Ann Arbor: University of Michigan Press

Taylor, J. 1998. 'Image of Contradiction: Obstetrical Ultrasound in American Culture', in H. Ragone and S. Franklin, eds. *Reproducing Reproduction: Kinship, Power and Technological Innovation*. Philadelphia: University of Pennsylvania Press, 15–45

Unnithan-Kumar, M. 1999. 'Households, Kinship and Access to Reproductive Healthcare among Rural Muslim Women in Jaipur', in *Economic and Political Weekly of India*. Mar 6–13, 621–30

—— 2001. 'Emotion, Agency and Access to Healthcare: women's experiences of reproduction in Jaipur', in S. Tremayne, ed. *Managing Reproductive Life: Crosscultural themes in fertility and sexuality*. Oxford: Berghahn, 27–52

—— 2002. 'Midwives among Others: Knowledge of Healing and the Politics of Emotions', in S. Rozario and G. Samuels, eds. *Daughters of Hariti: Birth and Female Healers in South and Southeast Asia*. London: Routledge

—— 2003a. 'Reproduction, Health, Rights: Connections and Disconnections', in R. Wilson and J. Mitchell, eds. *Human Rights in Global Perspective: Anthropological studies of rights, claims and entitlements*. London: Routledge, ASA series, 183–209

—— 2003b. 'Spirits of the Womb: Migration, Reproductive Choice and Healing in Rajasthan', in F. Osella and K. Gardner, eds. *Migration, Modernity and Social Transformation in South Asia, Special Issue of Contributions to Indian Sociology*. Delhi: Institute of Economic Growth, 163–89

Van der Veer, P. 1996. *Religious Nationalism: Hindus and Muslims in India*. Delhi: Oxford University Press

Van Hollen, C. 2002. 'Baby-friendly Hospitals and Bad Mothers: Manouvering Development in the PostPartum Period in Tamil Nadu, South India', in S. Rozario and G. Samuel, eds. *Daughters of Hariti: Childbirth and Healing in South and Southeast Asia*. London: Routledge

Visaria, L. and Ramachandran, V., eds. 2001. *The Community Needs-Based Reproductive and Child Health in India: Progress and Constraints*. Jaipur: Health Watch Trust

—— and Visaria, P. 2001. 'HealthWatch Case Study: Rajasthan and Tamil Nadu', in L. Visaria and V. Ramachandran, eds. *The Community Needs Based Reproductive and Child Health in India*. Jaipur: HealthWatch Trust

CHAPTER 4

PROGRAMMES OF GAMETE DONATION: STRATEGIES IN (PRIVATE) CLINICS OF ASSISTED CONCEPTION

Monica M. E. Bonaccorso

Extracts from Clinicians' Interviews

Clinician Rosso:[1]

'We say to the patients what we believe is right to say ... we do inform the patients in order for them to make a good decision ... but it is our duty (*dovere*) to judge every single case and explain to couples what they can understand ... it would be unfair to present to couples a series of options they cannot really understand ... couples would feel puzzled and it would not help'.

Clinician Bianchi:

'We make 70 percent of patient's decisions [the same clinician a few months later corrects the percentage and suggests 90 percent]'.

Clinician Verdi:

'The issue of information is a false issue in assisted conception ... it's a utopia ... of course this is something you should not write ...'[2]

Private Clinics of Assisted Conception

Without Legislation

In Italy the provision of infertility treatment has (at the moment of writing) a peculiar form. There is no legislation, but only an administrative act

– a *Circolare* (named after Degan, the minister who passed the regulation in 1985) – which establishes a general rule. The public sector is permitted to provide fertility treatment, in vitro fertilisation (IVF) and artificial insemination (AI), only if a couple's own gametes (egg and sperm) are used. The private sector can offer the same as well as treatment with third parties' gametes. The private sector can thus offer IVF with egg and sperm donation and AI by donor. Italy is an anomaly in the European panorama. In Europe similar programmes are regulated and not entirely privatised (for a general overview of legislation in Europe, see the Report on Bioethics of the European Parliament 1992; see also Gunning and English 1993 and Nielsen 1996. For specific examples see Novaes 1986 for the French case and Morgan 1993 for the British case).[3]

The Italian case is peculiar for at least three reasons. First of all, private clinics of assisted conception enjoy deregulation and full monopoly of gamete donation practices (e.g., selection of donors, collection and storage of gametes, number of donations per donor) not because of a voted bill but because of a lack of it. Secondly, programmes of gamete donation – widely perceived as the most problematic and controversial – are, more than anywhere else in Europe, in the hands of the private sector. They are fully located in the market place and pushed into the arena of commercial services; they are managed as a business by the clinics.[4] Thirdly, there is no audit. There is no *super-partes* authority entitled to check the services provided by the clinics. Clinical practice is not routinely inspected, unless a case of malpractice is reported and the authorities (the Nas, a special police body) are called in for investigation. Again this is an anomaly within the European context. Morgan, for instance, explains that in the British case the Human Fertilisation and Embryology Authority 'as [a] licensing body ... has power over public and private institutions to scrutinise and license, to approve, discipline and sanction the provision of assisted conception services and three main areas of activity; the storage of gametes and embryos, research on human embryos and *any infertility treatment which involves the use of either donated gametes* or embryos created outside the human body' (1993: 80, emphasis added).

The lack of a prescriptive legal framework and the status of medical services 'out there' in the market place stimulates specific representations. Private clinics use the rhetoric of personal choice to sell programmes of gamete donation, arguing that demand justifies private provision. But this is not necessarily the case as private clinics are the only place where programmes can be provided: there is no choice. The Circolare Degan forbid the public sector from offering programmes of gamete donation. If couples wish to proceed with programmes involving third parties they cannot but proceed with the provision of private services. Strathern makes a useful point about the rhetoric of commercial services: 'those who seek assistance, we are told, are better thought of *not* as the disabled seeking alleviation or the sick seeking remedy – analogies that also come to mind – but as *customers seeking services*' (1990: 5, emphasis added).

As such, couples – or customers – proceed with information about programmes of gamete donation gained and delivered by the private sector only. Information is available there and nowhere else.[5] It is given from 'within', during (private) medical consultation. It is there that couples are told of treatment with donated gametes, rates of success, screening/selection of donors and costs. It is there that couples start treatment. And it is there – in the space of a private clinic – that programmes of gamete donation are constructed as *the solution* for overcoming infertility. In a private clinic medical concepts are reshaped and become *flexible*. Infertility is transformed from being an irreversible disease into being a temporary deficiency. Of course what changes is not the diagnosis, but how definitive that diagnosis is. The mere option of a programme of gamete donation changes the 'status' of infertility. The new status is fictional. It is the fiction of the private.

Life in the Clinic

The clinic is the place where couples go for infertility treatment. There, conception becomes an explicit, mediated, non-intimate act. It is dislocated in the unfamiliar, interceded by medical practitioners and medical expertise, aided by unknown donors. The clinic soon embodies special relations between different people, who play a part in the procreative project. New alliances between providers (clinicians) and recipients (patients) come into existence (Price 1992) together with particular relations of power.

Clinicians are not just those who make the diagnosis, advise what programme to take on, provide (and choose) the gametes, disclose rates of success and failure, and decide costs. In the eyes of couples, they do the impossible. Clinicians hold an exceptional expertise with which they control the procreative project and make it possible. It is because of this expertise that they are fully trusted, and fully trusted is the clinic in which they operate. Couples invest heavily in them and have unconditional faith.

Cinzia says:

> Doctors can eventually make this happen. They can do what you cannot by yourself ... I cannot but trust and have faith in the miracles of modern medicine.

Trust and faith are capital in programmes of gamete donation and certainly in the private sector. They inhibit criticism or scepticism. Again in the eyes of couples, what clinicians do or say is always right. Clinicians ultimate desire is to offer couples the 'gift of a child' (*il dono di un bambino*) or 'transform the dream of a child in reality' (*fare del sogno di un bambino realtà*). This is what couples say and always seem to believe.[6]

Emma explains:

> The doctor told me to trust him (*avere fiducia*) and have faith (*avere fede*) ... once I was so depressed that I could not be examined ... then he said to lie

down and calm down that he knew what he was doing and promised I would have a baby soon or later ... I felt a sense of relief ... I often think that if he did not say that, in that moment, I would have left and never gone back ... I was very tired at that point.

Through faith and trust *dependency* is created. It is almost tangible: couples' lives revolves around the clinic. Couples do what clinicians suggest, scrupulously following every indication, engaging in new examinations, tests and treatment if so advised. No one seems to ever wonder if things are really all right.
Letizia says:

> I do what the doctor tells me to do. She knows and she wants to help us. I never object to an examination or anything else ... I told her to do whatever is necessary to have the baby ... I told her we do not look at the expenses ...

It is the state of mind to be one of dependency. Couples continually hold memories of events that have happened during consultation, or examination, or treatment. Wife and husband discuss and re-discuss every single event again and again, the way things went in one or the other occasion. Almost always they can recollect (literally) what the doctor did, or said, or explained, or promised. Couples keep it all in mind. Or, as Stefania explains below, 'in the head' (*tutto nella testa*):

> It all happens in your head, if you see what I mean ... to go to the clinic, and see the doctors, is the most important thing for me now ... it comes before anything else ... I think all the time about this place, what happens here, what they say to me ...!

Barbara says:

> It works in a way that you are always here ... even when you are somewhere else ... either because of practical matters or because it happens that you think about it ... may be you are doing something completely different and it comes to your mind ... or watch a film and start rethinking this or that ...

Giancarlo adds:

> It is an experience that absorbs me completely ...

Couples entertain continuous *imagined* relations with providers of treatment. As others, Elisabetta likes to think that the doctor cares about her. Even though, as she points out, the doctor does not say it.[7]
Elisabetta says:

> 'I think that by now the doctor cares about me. She knows me well now, always understands my mood ... sometimes I feel like she is sorry that she cannot do more for me. ... I feel that even if she does not say it I come into her mind as she comes to mine.

A Programme is (Almost) Forever

The anxiety and pressure that accompanies a programme of gamete donation traps couples into the medical 'machine' for a time that often seems forever. In eighteen months fieldwork, and in my subsequent correspondence with some of the couples interviewed, I was never informed of couples dropping out. Couples enter programmes of gamete donation and incessantly keep engaging in new programmes regardless of high costs and failures. As Franklin points out for the British case 'women and couples *come to feel they have to try IVF*' (1997: 14, emphasis added). They try and try again, and the more they try and engage with *the idea of* assisted conception, the less able to leave they seem to be.

Vittorio explains it:

We discovered my wife could not have her own child, felt an intense pain, and said to each other let's do immediately all we can do, afterwards we will think about the rest ... we met the doctor who said that he could help us ... we had no choice but accept ... we are still in his hands ... and this was three years ago now ...

Costanza says again:

We spoke for what seemed a long time ... but it was from Wednesday to Sunday ... the Monday after I called my doctor [a sort of GP] and asked for the name of a specialist ... two days after we met the specialist in a private clinic. Since then I have had nine artificial insemination [three cycles in two different clinics] and, in the last two years, three IVF treatments ... [in the third clinic]

Couples go through an impressive number of treatments/programmes. Unsuccessful treatment does not lead to 'pause', but conversely to 'action'. Clinicians always re-engage couples at the end of an unsuccessful treatment, and when that is no longer feasible, clinicians themselves suggest couples try another clinic. This is a system that secures a continuous recycling of couples. Couples move from clinic to clinic in the hope that (in the new clinic) the programme will work better. Each time the couple visits a clinic, the clinician signs the couple into a programme – regardless of medical history and the number of unsuccessful attempts already experienced.[8] Each time, the couple starts all over again – consultations, tests, examinations. Each time, a similar diagnosis is given and more or less similar information about programmes of gamete donation is made available. A clinician *openly* explains why clinics always engage couples in new programmes:

I respect the work of my colleagues, we all work in the same field and have the same objective: to give these couples the joy of a pregnancy, of a child ... however we do not have the same expertise, experience and surgical skills, we all have received different training ... I was trained in the States and in Australia ... I am a veteran in the field, internationally known ... my team includes English and American gynaecologists and embryologists ... I almost find difficult to keep track of my publication records [and laughs] ... when a couple comes and sees me I look at the medical history but disregard it ... seven out of ten times I can achieve what other colleagues have not been able to ... sometimes I feel sorry these couples had to spend all that money, if they came to me immediately many of them would already have a child, if not two or three ... I respect my colleagues but cannot say no to couples when I am almost always certain that I'll give a child to them ... it's a moral duty to take care of people who need medical help ... it is my duty a clinician ...

I then ask what happens in those three out of ten cases who are still unsuccessful – the question is asked much later on during the interview:

Well ... in those cases I do not destroy the hope of a child, couples who suffer from infertility are very fragile, I suggest they go and see another clinic ... sometimes changing the clinic helps ... couples often respond well ...

The system, or attitude, creates a pattern of repetition. To put it in the words of an Italian clinician who does not offer private infertility services 'couples looking for treatment are *coppie itinerani*' – itinerant couples. It is the frantic search for the 'solution' that obscures any other aspect, including the 'nature' of the relationship with the clinicians as providers of private services for infertility treatment.

The Medicalised *Couple*

With a programme of gamete donation the status of couples changes dramatically. Couples turn into patients. In the clinic they are treated and act as if fighting a 'disease' that can be eliminated. The body is ill and needs to be cured.

Margherita explains:

'Really it depends on the way you put it ... and on the way the doctor supports you ... when you start, it's a strange thing: on the one side it is more difficult than you thought ... when the doctor puts that syringe inside it's really weird ... on the other side it is really a medical thing: you take drugs, come to the clinic, have scans, it is like when you are ill and need to sort yourself out ... [infertility] is not really an illness, but it looks like one ... also because of the situation, you are on the gynaecological bed ... in a kind of surgery room, the doctor speaks to you in a certain way and everything is medical ... you cannot *really* say you are making a baby ... it does not sound an extreme thing to do ...'

The medicalisation of the couple is a powerful strategy in programmes of gamete donation.[9] It helps clinicians to normalise treatment (see Cussins 1998a, 1998b). If the couple is *ill* clinicians can continually establish reasons for medical intervention. If the reproductive body does not work, and requires 'therapies', then the couple has a medical reason to attend the clinic. It is the *medical reason* that makes the programme an acceptable option. As Noveas writes: 'the medicalisation of AID appears as an attempt to create a socially validating framework for this practice' (1986: 579, see also 1998). It is because of the rhetoric of a medical need that no one in a clinic of assisted conception seems to ever wonder if the extreme medical invasiveness is really necessary or justifiable. No one there wonders if it is justifiable to undergo treatment after treatment, take drugs to super-ovulate, undergo surgery to extract the eggs and to have anaesthesia against pain. Medicalisation conceals the lack of limits or perspective on treatment itself – in a sense the cultural motives that make technologies of procreation sought for. Culture is medicalised to reproduce culture.

Medicalisation also displaces two other interdependent features. First of all infertility. A programme of gamete donation inherently confirms the impossibility of a therapeutic intervention in the strict sense; if the couple has to rely on gamete donation, it means that nothing can be done to reinstate fertility. What the programme does is *to add in* a third party to achieve conception. As Strathern points out 'the new techniques of fertilisation do not remedy fertility as such, but childlessness. *They enable a potential parent to have access to the fertility of others*' (1990: 6, emphasis added). Medicalisation thus serves the purpose of obscuring what a programme of gamete donation does; that is to fictionalise the reproductive event. Secondly, medicalisation obscures the replacement of couples' genetic material. The extreme emphasis on medical aspects and procedures helps to distance third parties. It turns the attention away from ideas of donors, acts of donation and bodily emissions. Medicalisation gives couples a different focal point. As Gioia explains the programme fades 'the worst bit of it' by medicalising the intervention and emphasising the benefit of being '100 percent under medical control'.

Gioia says:

> It helps that things are organised in a way that, yes, on the one hand you get all that comes with it ... which is really the worst bit [donation of sperm] ... but on the other hand it is also a way to take care of yourself and your body in a medical way ... I am going through lots of tests and exams which is good because it helps to prevent other diseases ... I am 100 percent under control and my husband too ... so it is good. It's like having a doctor in the family! If you understand what I mean.[10]

The Technological Body

High medicalisation results in the creation of a 'technological body'. Clinicians put extreme emphasis on the ill/unreproductive body and ask couples to do the same. The body needs constant attention; it needs to be pushed into the technologies. Couples are asked to think of and treat their bodies as extensions of the technologies if they wish for (they are told) the technologies to succeed. Valentina explains it:

> One has to become very efficient. Your body has to respond to the therapy, doctors ask you to take pills, or have injections, every day at the same time ... to go through treatment is to make it explicit that your body has

failed, you have failed ... that's why you put all yourself into it ... into the programme and feel okay about the drugs and all the rest of it ... you want to make it work for once ... you think about it as something detached from you ... but still part of you ... the doctors give the orders, you want it [the body] to listen to the doctors and to you too ...

Mariangela problematises the focus on the body, the high level of technological intervention; but, as do many other women and couples, believes that this is the way it needs to be for the programme to work. Mariangela feels that she has lost 'face and soul':

You feel as if you are nothing but a uterus and ovaries. All the attention is there, doctors look at there, not at you ... you have no face or soul (*anima*). You have to respond ... if your body does not respond to the technologies you are responsible ... the doctors cannot do what they are supposed to and the technologies will not work ...

Francesca says:

It's a learning process if you wish, the doctors teach you what to do with your body, how to prepare it for the technologies ... you learn it and you do it. Quite soon, I started to see my body as a machine that has to accept technology ... by itself my body is nothing, it is dead ... the technology does the work that my body does not, yet it [the technology] needs my body ... what is left in it ... this is how the doctor explained it to me ...

Clinician Rosa says: 'Technology is aid, and as any aid it needs the participation of the aided.' Technological newness in the making of babies is used to stress the inefficiency of a body, which is not simply unreproductive, but often unwilling to welcome/receive technology. The emphasis on the bodies of couples is extreme. It is phenomenal the way in which, from being an intervention that helps couples, technologies of procreation turn into interventions to be aided by couples. Technology thus stands, at once, for both aid and its reverse, progression and arrest.

The Language of the Assisted Conception

The language of assisted conception is made of continuous oppositions. It operates in a binary way that always incorporates one thought and its opposite – yet without marking the contradiction of contrasting statements. Explicitness always implies obscurity and mystery, knowledge always entails ignorance, awareness always hints at unawareness. The particular language serves to manage the provision of treatment: it assists the technologies. It is also, of course, more than just language, it is the culture of the technologies. As Bourdieu (1992) says, the exchange in words is an exchange in meanings.

Clinicians, embryologists and biologists (all those working in a clinic) present the technologies to the world outside (the 'public space') and to the world 'inside' (the clinic) in different ways. In the world outside, the mediascape (Ginsburg 1994) or any public/official settings, they employ a strategy of *explicitness*. They use a public language. Programmes of assisted conception are revealed; they are presented to the general public in very immediate ways. Visual and textual narratives offer detailed accounts, and seem to portray the complex reality of the technologies. Clinicians spell out implications, risks, failure rates. They often accurately inform on some distressing implications.[11]

By contrast, in the clinic and in everyday interaction with couples, clinicians operate a strategy of *non-explicitness*. Or decency, as one clinician put it to me. Within the walls of a clinic a clinician is no longer just an 'expert', a 'scientist', but also a (private) provider of infertility services. In the clinic clinicians sell the technologies. This is why a number of strategies need to be put in place, and why they have some powerful internal contradictions.

Managing Failure

In the daily routine of an infertility clinic the reproductive technologies are said to be therapies more or less equivalent to other medical interventions. A programme, as a 'therapeutic' enterprise, needs to be normalised, assimilated to medical routine. In everyday practice couples cannot be continually reminded that the technologies are experimental, risky and probably unsuccessful. This is how clinician Viola explained it to me whilst scanning the ovaries of a patient:

> The NRTs (New Reproductive Technologies) are therapies which favour reproduction as much as contraception obstructs it ... the NRTs are nothing more and nothing less than the result of certain applications in modern medicine ... if you accept in principle human intervention on the human body, thus medical intervention on the ill body, you will have to agree with me that you cannot but accept fertility treatment as a therapy for the cure of infertility.

However, as soon as failure occurs the tale is complicated.[12] The general perception of the technologies is not effected, *but only the experience of the couple who is encountering failure*. Suddenly, with respect to that specific treatment, delivered to that specific couple, the technologies are no more therapies. Contrasting ideas are put forward to address different, opposite and simultaneous reasons depending on the case or moment or, sadly, even the couple. In any case, inconsistent as it may seem, failure is soon linked to the couple undergoing treatment. When failure begins to occur, the couple begins to be invested of responsibility – and guilt.

In the context of a treatment that involves a specific couple – the possibilities, chances, outcomes of medical intervention became plastic. As soon as the programme proves to be unsuccessful – perhaps again and again (?!) – the technologies become increasingly exceptional medical procedures. They do, according to the clinicians, because of what they call *the specific couple*. It is the couple that turns the technologies into one thing or the other. It is thus the couple that makes the technologies work or not. Not the other way round.

Clinician Viola explained it again (in the course of a consultation with a couple who repeatedly failed):

> We explain to couples that these are complex technologies, you see, the latest advances in the science of reproduction; when we incur an unsuccessful treatment we explain that the same result cannot always be achieved.

A few minutes later, still during consultation (the couple is sitting opposite to us) clinician Viola explains again:

> This particular couple for instance has a negative response to the procedure. I have already discussed it with both of them, did I not? [Viola smiles to both wife and husband]. *It happens sometimes, not that often, but you see we are all different and unconsciously may not accept the treatment.*

Failure is never related to medical practice. Never does a clinician take responsibility for it. Never does the clinician suggest the possibility that the quality of donated gametes (including collection and conservation measures), the standard of procedures (manipulation of gametes and conservation of embryos), or technological expertise (surgical skills for instance) might have played a role during treatment. The basic fact that *all* these elements interfere and contribute to success/failure is consistently, and deliberately, disguised.[13]

Clinician Rosso explains:

> We *touch* extremely complex mechanisms, we enter into a space dominated by the laws of mother nature ... we above all deal with the personality of many different couples, their physical and psychological response ... *often it is the patient which responds in an unpredictable way.*

Thus, in the words of the clinicians, unpredictability rests (not so much) in the element of technological sophistication but (and especially) in the couple. It is for the couple – materialised in the person undertaking treatment – to respond to the technology. The couple can respond well or sometimes exceptionally, but often does not. It fails. As another clinician puts it 'not everybody responds adequately, yet'. 'Yet', she says, perhaps implying that if the couple were in the future to respond more efficiently, medical technological intervention will be too. The couple – or the person in the couple – is faulty, not the technologies! The more the technology is advanced, the more failure is encountered, the more the couple is inept.

Understanding the Technologies

The language of assisted conception is shaped into the *form* that information and knowledge take in the clinic. There are two systems of communication in place. A system that exposes couples to a highly sophisticated medical and technological language – in Rowland's words 'reprospeak' (1992) – and a system that relies heavily on a language that I call the language of *commonplaces*. It is precisely the distance between the two which illuminates the artificiality of the device.

Clinicians have a strong interest in using both, in parallel and in sequence. The ability to use one or the other, at the right time and circumstance, makes a clinician successful – that is with a crowded waiting room and a long waiting list of willing future clients. The skill rests in the capacity to formalise/deformalise descriptive and explanatory medical procedures to suit each situation; it also rests in the facility with which a clinician shifts from one to the other. Descriptive and explanatory formalisation is introduced when 'distance' with the couple needs to be taken. A sophisticated medical language is adopted to secure control over the couple through linguistic jargon and hypermedical concepts. Deformalisation is introduced when 'closeness' with the couple becomes necessary. In this case a language of commonplaces is again adopted to secure clinician control over the couple. The clinician simultaneously reassures the couple and blocks off the amount and type of information.[14]

How are the two languages put to work? The formalised, hypermedical language is used when the couple is challenging and willing to gain more information and knowledge. The couple is openly questioning the treatment, is asking for specific detail about recruitment, selection and screening of donors. The hypermedical language is used to convey, and partly display, expertise – the clinician knows best, knows more than anyone else, the clinician should not be questioned, too specific questions should not be asked. With the hypermedical jargon a deliberate attempt is made to intimidate the couple – the more the couple shows interest and ability/willingness to apprehend, the more the clinician inflates the jargon to widen the distance. The deformalised language of commonplaces is used when the couple wants to gain more information and knowledge but shows confusion, is not direct or does not openly question treatment; the couple finds it very difficult to ask – and phrase – questions. A language of commonplaces is used to convey reassurance, and partly patronise the relationship with the couple. This is a language that triggers emotional responses. With the jargon of commonplaces the clinician shows empathy. S/he reinforces the relationship with the couple and leads it.

With both languages the final outcome is the same. The clinician is empowered against an increasingly disempowered couple – the longer the treatment is, the more the couple is disempowerd. In contrast to what one may suspect, the use of one or the other terminology is not directly dependent on the 'class' or 'education' of the couple. The ability to form and articulate questions is, for instance, not an obvious reason for employing an hypermedical language. It plays a part in it, but there are other factors that the clinician considers too. The clinician, as I said earlier, is proportionally successful to his/her ability to understand at a very early stage what *touches* a particular couple. The clinician uses the language that serves her/his interests best.

Clinician Giallo gives his own explanations of the two systems of communication in place. He says:

> There are patients who have an *instinct* for medicine, they grasp what you are talking about ... with them I can use the medical language and they will understand. There are others that need to be supported, explained what it is all about ... with those you need to use different words ... it's a bit like with my children ... if I have to explain how the heart works, I explain it in very simple terms ... I use the example of a pump that pushes air inside the wheel of a bicycle ... the heart works more or less in the same way ... the heart is a pump. This is what I do with my patients ... I make them understand what they need to know ... but you see it is not really essential to know everything in life ... It is also okay not to know that much.

I then ask how he explains to 'patients', especially those who in his view do not have an 'instinct for the medical language', the genetic implications of using a donor for instance. He replies:

> I tell them not to worry, that children are always children, and it does not really matter the genetic component ... I tell them that we make sure we provide good donors, and we screen them and look for some similar physical characteristics ... I reassure them ... I tell them that I have two children and they are very different ... you would doubt that they are brother and sister.[15] I take it philosophically ... [16]

It is interesting however, that although clinician Giallo's explanation differs from mine, his words implicitly confirm that in the clinic there are two parallel systems of communication in place and that these are managed by the clinicians. It is interesting because when I asked if this was the case, he replied with a definite no.

The Domain of Kinship

Clinicians use and exploit certain ideas, concepts and, even beliefs, which will facilitate, propitiate, favour technologies of procreation. They use specific *words* to evoke familiar ideas (the same ideas that, in the first place, inform couples' choice and acceptance of a programme of gamete donation). In particular, clinicians draw upon the known domain of kinship to rationalise/sustain the technologies. As Ragoné points out for the U.S. the language of the NRTs is fundamentally a kinship language. It is used in the clinic from the very beginning of treatment: '... kinship terminology is routinely employed even prior to the conception or birth of the child' (1994: 39). The language of kinship helps couples to become familiar with the technologies. It facilitates the integration of the technology itself. It is employed to direct, sustain and motivate couples at any stage of the process. It helps to overcome doubts and fears. It helps to validate the attempt, the effort and the cost of a programme of gamete donation.

The language of kinship reiterates specific ideas about infertility as negative and a deeply 'unnatural' condition; the desire for children and so the desire to reproduce oneself as fundamentally 'natural'. Furthermore, that having a family is the most desirable, if not compelling, achievement in one's life, and the biological and social relatedness between kin as indisputably significant. It is precisely because these ideas are incessantly reiterated that the language of kinship works, and clinicians heavily deploy it. The language of kinship is a strategy that helps clinicians (and couples too) to keep going with programme after programme.

Clinician Marrone says:

> When a couple come and see me I explain what I can offer, and what I cannot. I immediately say that I cannot make miracles, but I also say that I may be able to give them what they are looking for ... *a baby, a family, the joy of a birth. I explain to my couples that if we are lucky they may hold their baby soon* ... with three or four cycles of artificial insemination by donor they can have a baby ... in all those other cases in which the patient needs to undergo an IVF cycle the percentage changes, but it depends from case to case. I help them to believe that this is going to happen, a positive psychological approach is what they need ... *they need to believe that they, as every body else, will become parents.*

Clinician Marrone, as many others (with almost no distinction between men and women clinicians), uses a very familiar/kinship language to state his intention to give couples what they are looking for : '... *a baby, a family, the joy of a birth. I explain to my couples that if we are lucky they may hold their baby soon* ...' he says. Clinician Marrone, as others, knows that this language, this particular conceptual or cultural language, will tranquillise every couple – every couple who desperately seems to want/need to have a child, who incessantly attends the clinic and engages in one after another programme of gamete donation.

In the same fashion are framed ideas of donation as a safe practice for family life, the donation of gametes as an act without moral and familial implications, donation as a solution which will sooner or later, bring happiness amongst all kin.

Clinician Nero says:

> Couples need to understand that gamete donation is Okay, that *it does not really create complications in family life*. When the baby comes couples forget about the donation, I know this by experience. I always tell my couples about the experience of other couples, who went through it before ... there is nothing better than first hand experience. I want to make them understand that we have good statistics, *when the baby arrives, when the whole family sees the baby, they will forget.*

Clinician Arancione explains:

> ... I have always a preliminary meeting with the couple, to make sure that they are Okay, that there is no problem with the idea of donation ... I

explain that it all may look different because they come here, in the clinic, to have their baby, but it is really not that different, *the donation of gametes does not change their being parents*. In a sense it is a technicality ...

Ideas of biological substances are again conceptualised in the same fashion. In the course of a programme, donated gametes are emptied of their properties and uniqueness. The emphasis is put elsewhere – the significance of the genetic tie between kin is ignored and replaced with the significance of the social and emotional tie. The shift significantly overrides the extreme emphasis on genetic ties used to engage couples in the first place (e.g., programmes of gamete donation are always presented and justified with the idea that at least 50 percent of a genetic tie between the fertile parent and child is maintained).[17]

Barbara says:

> I was worried because my child will be different from me but the doctor said that *what makes a child is the way you bring him up* not really the genes or anything like that ... then I asked what about being sure to have a clever child ... she said that nowadays science has proved that *it all depends from the environment and from the stimulation parents give to children ...*'

Clinician Blu explains how he suggests to couples the 'right' (less problematic) way to think about gamete donation, gametes and the making of persons. He says:

> I am not only a doctor here, but also a kind of psychologist: *I always explain to couples that what makes a person is the social environment ... It is important to be good parents, to be good mothers and good fathers* ... genetic inheritance plays a minimal role ... I always point out to them the case of two twins who are separated at birth and live with different parents ... they will be different. *It is the family that matters, the parents and the relatives* ...

It is because of the language of kinship, the ambiguous use of that language, that couples are sustained and sustain themselves through the process. The language of kinship is used to 'naturalise' programmes of gamete donation. Notions of naturalness are always associated to notions of kinship. The language is special not in content, but in its capacity to evoke associations. As Bourdieu (1992) suggests, the exchange of words is an exchange in meanings. It is a language of very *selected* notions easy to reproduce. It is a language that reproduces known patterns of questions and answers. It is very predictable. It is this predictability that makes it an excellent medium in clinics of assisted conception.

Conclusion

Clinics of assisted conception are an extraordinary microcosm. The minimalist and almost aseptic physical space contrasts with the complexity with which *clinical practice* is constructed. Specific cultural and social constructs circulate in the clinic and are deployed when necessary. They serve a double role: on the one side clinicians use them to convey to couples certain *messages* (which carry certain meanings) in the course of programmes of gamete donation. That is to say that couples in buying the technologies also buy a way of thinking about them.[18] On the other side, the same constructs help clinicians to elaborate a narrative for themselves. Clinicians live in the social and cultural world they attempt to manipulate. They also need to fully believe in what they do.

This is novel material. To date the field of assisted conception has been entirely dominated by a Euro-American perspective. We have accumulated anthropological knowledge of the NRTs as they take place predominantly in North Europe and the U.S.; in cultural contexts (anthropologically) distant from South Europe. The material I have presented here, contrary to a widespread presumption within anthroplogy, shows that there is nothing so inherently and peculiarly Mediterranean. Rather, it shows some interesting continuities with the existing literature on the NRTs, and particularly with American works. The 'culture' of the NRTs, as it emerges in Italian clinics of assisted conception, has intriguing points of connections with the 'culture' of the NRTs as it emerges in the works, for instance, on American surrogacy programmes (Ragoné 1994) and American clinics of assisted conception (Cussins 1998a). The constructiveness of the NRTs makes itself equally explicit in Italy as in the U.S. The common denominator being a free market and the private provision of infertility services.

To clarify the point in relation to the Italian case; the lack of legislation – which also means full monopoly and absence of audit – gives clinicians (and generally medical staff) full freedom to operate as they please. The absence of legally binding regulations opens up a space for adopting the best, most suitable and convenient, strategy of action. Of course this is a *cultural* strategy. It makes it possible to extensively manipulate those social and cultural constructs that facilitate the interaction with couples, especially kinship constructs. It makes it possible to excessively justify the aid of the technologies and secure a continuous demand for the clinic and the provision of services. In contrast to other European contexts, where prescriptive legal frameworks are in place (such as in the U.K. for instance), the combination of the lack of legislation and a free market permits a closer investigation of the clinical body. I do not argue that the lack of regulation makes the Italian situation unique, rather that it puts into greater relief the types of discourse that are used throughout Europe and the U.S. It facilitates observations of the social and cultural constructs that sustain the use of technologies of procreation.

Notes

1. These are fictional names. I use colours.
2. I quote what I was asked not to because I wish to render the duality of medical discourses. It is this duality that informs treatment and it is from this duality that clinicians talked to me – always pointing out what I should say and what I should not. And always offerings two versions, in the form of two different statements: the official (which they always asked to write about, 'questo lo scriva, mi raccomando') and the confidential (which they asked not to write about, 'questo non lo scriva, mi raccomando').
3. The reasons why Italy lacks legislation are quite complex, and beyond the scope of this paper (for a full analysis see Bonaccorso 2000). In brief: for the last twenty years, programmes of gamete donation, and more widely the health/reproductive technologies have been used instrumentally to assert political oppositions. These technologies offer tremendous opportunities for political manipulation. Both left and right wing parties (at the moment of writing there are 52!) have been heavily exploiting such an opportunity.
4. In Italy no other medical treatment (linked to a pathology) has the unique status of being provided by the private sector only. This gives to programmes of gamete donation an unusual, unique, status; to follow Morgan (1998) it contributes to enlarge the global market of assisted conception services.
5. The public sector does not provide the information, as it cannot provide the treatment, and there is no other independent source. There are a number of organisations, associations of patients, which have the mandate to inform couples in need. These are often financially supported by clinicians (who of course are the same individuals owning private clinics of assisted conception). This makes information non-independent. It is worthwhile mentioning how peculiar the lack of a neutral informative space is in the Italian context. Independent information is generally common in the most disparate fields and it is usually influential.
6. Unconditional faith – which is often synonymous of lack of criticism – is especially significant in the Italian case. Italians have a 'cultural attitude' towards criticism. This is not an oversimplification, though a generalisation. In particular, the relationship between the public and the private health sector is highly politicised. It is interesting that the private provision of infertility services is not.
7. Of course clinicians do not seem to be 'thinking' about their couples, but always pass on subtle messages of personal involvement with the specific case. This stimulates in couples emotional responses.
8. During fieldwork I never saw, in any of the five clinics where I worked, a clinician refusing to provide treatment.
9. This is a strategy within the clinics.
10. The comment is particularly interesting if I add that at the beginning of the interview, and in a different conversational context, Gioia explained that she is a 'naturalist' (*una naturalista*).
11. In order to capture the language of assisted conception as it is spoken in a clinic, it is helpful to keep in mind the contrast with the language used publicly. I suggest the possibility of a public space, a language spoken publicly and a general public, following Morely's argument that global and globalised events reach via the media the privacy of our lives (1992).

My claim that experts problematise publicly their work should not come as a surprise. News is constructed with information collected from 'scientists'. The criticism/problematisation that often emerges in media account of the NRTs does not necessarily represent the critical view of the producer/journalist. The 'experts' themselves often offer a problematic picture of the treatment they provide. (For a more detailed account of the way in which the technologies are debated publicly see again Bonaccorso 2000).

12. Of course failure is one, if not the main, problem clinicians have to cope with day by day, and with every couple (contrary, as it will soon become clear, to what couples are made to believe).
13. The medical literature on the subject is vast. For a general overview see Medline.
14. There are a number of points to be made here. The first one is that the strategy of a highly medicalised language is possible because – whilst couples came from all walks of life – they are very rarely medics. This is not a mere coincidence but a fact itself – none of the clinicians, embryologists, biologists interviewed ever responded that they would themselves rely on programmes of gamete donation. Overwhelming couples with a hypermedical language works precisely because the recipients are not medics. The second point, I as a social anthropologist doing research in the clinic experienced it: clinicians would often use hypermedical language when they wished to gain a position of power and obstruct my questioning. It was intimidating. Although I could understand, I found it difficult to reply and/or ask a new question using the same jargon. The lack of training as a medic makes it difficult, if not impossible, to sustain endlessly such conversations. Interestingly, in one of the clinics where I worked, such attitude faded when two clinicians realised that, although I was not a medic myself, I probably had sufficient familiarity with the terminology as the daughter of a medic. From this moment onwards every question I asked was answered with another question that in principle should have stimulated an emotional response. An example: when, at the beginning of fieldwork, I asked if donors were really matched with couples, the clinician gave me an apparent detailed account of what he called 'the genetic matching'. When, towards the end of fieldwork, I asked the same question to the same clinician he said: 'let's leave out the field of genetics ... let me put it this way ... if you were suffering from infertility would you want to be picky about the donor or would you want to make sure you become pregnant?'
15. Clinician Giallo always omits to say that the two children were born from two different mothers. He is divorced and remarried. The majority of couples I talked to and were undergoing treatment with him would tell me the same story: '... the doctor explained that he has two children and they look so different anyway!'
16. This view did not just emerge in the interview with Clinician Giallo. Generally interviews with clinicians are full of assumptions about what people can understand or not. At first I thought clinicians were just particularly opinionated. It seemed they believed in a world of able people against a world of not able ones. Afterwards I realised, putting together the interviews with clinicians and couples, that these are not mere bias. These are devices that help clinicians in their daily routine – these views contribute to justify oneself the consistent omission of information.

17. This is an extremely important point. The emphasis on the genetic tie – the maintenance of at least a 50 percent genetic connection between the fertile parent-to-be and the offspring – is what seems to motivate *all couples* choosing programmes of gamete donation. It also continually emerges from clinicians offering the programmes. However, in due course, and when a third party's gametes materialise in the reproductive event, the significance of the genetic connections is totally minimised to permit the acceptance of third parties genetic aid.
18. It may clarify the point to say that the technologies, in order to be accepted, need to become intelligible for the couple.

References

Bonaccorso, M. 2000. *The Traffic in Kinship: Assisted Conception for Heterosexual and Lesbian and Gay Couples in Italy*. University of Cambridge, unpublished Ph.D. Dissertation

Bourdieu, P. 1992. *Language and Symbolic Power*. Cambridge: Polity Press

Cussins, C. 1998. 'Producing Reproduction: Techniques of Normalisation and Naturalisation in Infertility Clinics', in S. Franklin and H. Ragoné, eds. *Reproducing Reproduction: Kinship, Power, and Technological Innovation*. Philadelphia, 66–101

—— 1998a. 'Strategic Naturalizing: Kinship in an Infertility Clinic'. Unpublished Paper Presented at Symposium New Directions in Kinship Study

Daniels, K. and Haimes, E. 1998. *Donor Insemination: International Social Science Perspectives*. Cambridge

Edwards, J., Franklin, S., Hirsch, E., Price, F. and Strathern, M. 1999 (1993). *Technologies of Procreation: Kinship in the Age of Assisted Conception*. London: Routledge

Franklin, S. 1997. *Embodied Progress: A Cultural Account of Assisted Conception*. London

—— and Ragoné, H. 1998. *Reproducing Reproduction: Kinship, Power and Technological Innovation*. Philadelphia

Ginsburg, F. 1994. 'Culture/Media: A (Mild) Polemic'. *Anthropology Today*, 10: 2, 5–15

Gunning, J. and English, V. 1993. *Human in Vitro Fertilization: A Case Study in the Regulation of Medical Innovation*. Aldershot: Dartmouth

McNeil, M., Varcoe, I. and Yearley, S. 1990. *The New Reproductive Technologies*. London: Macmillan

Morely, D. 1992. *Television Audiences and Cultural Studies*. London: Routledge

Morgan, D. 1993. 'Undoing What Comes Naturally – Regulating Medically Assisted Families', in A. Bainham and D. Pearl, eds. *Frontiers of Family Law*. London

—— 1998. 'Frameworks of Analysis for Feminisms' Accounts of Reproductive Technology', in S. Sheldon and M. Thomson, eds. *Feminist Perspectives On Health Care Law*. London.

Nielsen, L. 1996. 'Procreative Tourism, Genetic Testing and the Law', in N. Lowe and G. Douglas, eds. *Families Across Frontiers*. The Netherlands

Novaes, B. S. 1986. 'Semen Banking and Artificial Insemination by Donor in France: Social and Medical Discourse'. *Journal of Technology Assessment in Health Care*, 2: 2, 219–29

—— 1989. 'Giving, Receiving, Repaying: Gamete Donors and Donors Policies in Reproductive Medicines'. *Journal of Technology Assessment in Health Care*, 5: 4, 639–57

—— 1998. 'The Medical Management of Donor Insemination', in K. Daniels and E. Haimes, eds. *Donor Insemination: International Social Science Perspectives*. Cambridge: Cambridge University Press, 105–30

Price, F. 1992. 'Having Triplets, Quads or Quins: Who Bears the Responsibility?', in M. Stacey. *Changing Human Reproduction*. London: Sage, 92–118

—— 1995. 'Conceiving Relations: Egg and Sperm Donation in Assisted Conception', in A. Bainham, D. Pearl and R. Pickford, eds. *Frontiers of Family Law*. Chichester: Wiley

Ragoné, H. 1994. *Surrogate Motherhhod: Conception in the Heart*. Boulder: Westview Press

Report on Bioethics of the European Parliament. 1992. Directorate General For Research. 1992. *Bioethics in Europe*. European Parliament

Rowland, R. 1992. *Living Laboratories: Women and Reproductive Technologies*. Bloomington: Indiana University Press

Strathern, M. 1990. 'Enterprising Kinship: Consumer Choice and the New Reproducive Technologies'. *Cambridge Anthropology*, 14: 1, 1–14

Stacey, M. 1992. *Changing Human Reproduction*. London: Sage

Stanworth, M. 1987. *Reproductive Technologies*. Oxford

CHAPTER 5

WOMEN, DOCTORS AND PAIN

William Stones

Using experience from Sri Lanka (Simpson, Chapter 2) and Italy (Bonaccorso, Chapter 4) the ways that modern fertility treatments are viewed by society at large, by patients and by clinicians have been discussed. The present writer, an obstetrician/gynaecologist working in the British health service, considers the ambivalence that arises when undertaking consultations with patients within a biomedical paradigm, but with an awareness that social and cultural undercurrents rather than disease processes actually determine patients' expectations and care seeking. Pain is often the prompt for medical consultation but also occurs as part of normal life experience. Sociocultural factors are critical in determining how an individual, either doctor or patient, identifies a particular pain as worthy of attention. This chapter first discusses the biomedical view of pain in terms of current neuroscience together with recent understanding of sex (as opposed to gender) differences in pain perception and its heritability. Medical agency is then considered in the context of pain treatments. Finally, the special experience of women is considered: pain is associated both with the normal processes of reproduction, as in menstrual cramps and childbirth, and with specific diseases of the reproductive system.

Pain

The first biomedical model of pain was that of Descartes, who illustrated the nervous impulse travelling from the site of tissue injury to generate the sensation of pain in the brain. The most influential conceptualisation of modern times is the Gate Theory of pain which emphasises the role of the dorsal horn of the spinal cord as the locus for modulation of pain by descending input from the central nervous system (Melzack and Wall 1965). Current neurophysiological models emphasise the nature of pain as a central nervous system construct, and reject the notion of separate pathways for touch and pain. In this evolved model it is not possible to compartmentalise psychological and sociocultural as distinct from biological aspects of pain (Berkley and Hubscher 1995). It is not that pain exists and is then subject to psychological or sociocultural influences, but rather that pain is a product of the entire conscious organism functioning under certain conditions. This conceptualisation accords with the observation from clinical practice with chronic pain patients, that unimodal treatments for pain, especially those directed towards specific 'pain pathways', are unlikely to be effective.

The current IASP (International Association for the Study of Pain) definition of pain is 'An unpleasant sensory and emotional experience associated with actual or potential tissue damage, or described in terms of such damage' (IASP Task Force on Taxonomy 1994). This definition emphasises the subjective nature of pain; in accordance with the evolved view of pain described above, reference to 'pain pathways' is inappropriate as is the notion that specific parts of the nervous system convey perceptions such as touch or pain. Rather, the elements of the nervous system, comprising the 'different fibres, tracts, pathways and nuclei process and convey information about bodily stimulus events' (Berkeley and Hubscher 1995). While it is not possible to measure 'objective' pain an extensive literature has arisen describing valid and reliable pain assessment measures for humans such as visual analogue scales (VAS) and multidimensional questionnaires such as the McGill Pain Questionnaire. In animal experimental research on pain reproducible behavioural models for different types of pain have been developed which form the basis for drug development. There are of course questions about the applicability of animal experimental findings to human pain and certain agents while effective in some species are not effective in others.

Current neuroscience distinguishes *inflammatory* and *neuropathic* types of pain. In the first, pain arises as a result of the release of pain producing substances as a result of tissue injury. This injury may come about as a result of trauma, infection or non-infective inflammation such as rheumatoid arthritis. A second type of pain, neuropathic pain, arises when nerves are damaged and become excitable under inappropriate conditions. Examples include diabetic neuropathy and certain types of facial pain. A third type of pain, that arising from the viscera, is of especial interest to the present discussion. 'Viscus' refers to the soft internal organs of the body such as the bowel, bladder and uterus. *Visceral pain* cannot at present be classified clearly as inflammatory or neuropathic, although it has some of the features of both. Distinct features of pain arising from viscera include neurotransmission through slow-conducting C fibres, poor localisation to anatomical sites, and viscero–somatic (referred pain) and viscero–visceral convergence, where stimulation of one viscus such as the bowel causes measurable changes in another organ such as the bladder or uterus. (The sensory nerve supply to these 'visceral' organs is anatomically less complex and precise compared to that of the 'somatic' tissues such as skin and skeletal muscle).

In animal experimental studies, stimulation of the urinary bladder by distension under normal conditions excites only around 2.5 percent of fibres. Following the application of a noxious agent such as mustard oil previously silent fibres are excited, so that around four times as many fibres are recruited. This phenomenon has been described in studies of cutaneous thermal sensation but is less clearly seen where the mode of stimulation is mechanical. Given that the viscera are more liable to mechanical than thermal stimulation under normal conditions, it appears that the presence of numerous silent afferents and their sensitisation to mechanical stimulation is a characteristic feature of visceral as compared to somatic afferent innervation. This type of sensitisation may explain the common clinical observation of an acute episode of infection acting as the trigger to a chronic pain state, long after the original inflammation has subsided: previously silent nerves continue to discharge, and reconfiguration of sensory pathways within the spinal cord occurs.

An emerging literature identifies measurable influences of sex and gender on pain. There is a marked sex difference in the prevalence of a number of chronic painful conditions unrelated to the reproductive tract in the general population, such as irritable bowel syndrome, temporomandibular dysfunction and interstitial cystitis, which has prompted research on the potential underlying mechanisms. In studies of human volunteers, pain thresholds in women at different stages of the menstrual cycle using electrical stimulation showed that pain sensitivity was consistently lowest during the luteal phase of the cycle. In another study using pressure dolorimetry on the skin, pain thresholds did not vary during the menstrual cycle in women, but the number of points on the body tender to deep pressure was least during the luteal phase. No variation was seen in users of the oral contraceptive pill. Using thermal stimulation, Fillingim and colleagues did not find variation in sensitivity through the menstrual cycle but observed reduced sensitivity to ischaemic pain during the midfollicular phase of the cycle (Fillingim et al. 1997). These contrasting data indicate that the elucidation of the influence of hormonal factors on pain sensitivity requires a range of investigative modalities, a finding emphasised in the report of a meta-analytical review of pain perception across the menstrual cycle. This review estimated the effect size for menstrual cycle fluctuation of pain sensitivity between the most and least sensitive phase to be 0.40; this was contrasted with the effect size for sex difference of around 0.55, indicating that hormone variability could account for a substantial proportion but not all of the observed differences (Riley et al. 1999).

Evidence for a genetic basis for variation in response to painful stimulation and potentially in predisposition to chronic pain syndromes is emerging. In particular, quantitative trait locus studies using mouse strains have become feasible. Mogil and colleagues (1999) have investigated the heritability of nociception using twelve modalities of experimental pain measure applied to male mice of eleven inbred strains. They identified a range of genetic distinctiveness within the eleven strains, reflecting the strains' genealogical origins, and went on to analyse correlations indicative of types of nociception with a common underlying physiological basis. Three clusters of nociception were revealed, whose respective common features were:

1. Baseline thermal nociception
2. Spontaneously emitted responses to chemical stimuli
3. Baseline mechanical sensitivity and cutaneous hypersensitivity.

There is clearly a long way to go before conclusions clinically relevant to pain in humans can be drawn from the above data. However, there are already some important observations in this field. For example, some information is available from twin studies with regard to painful conditions affecting women. In an Australian cohort of female twins who were questioned on two occasions eight years apart, the longitudinally stable variance attributable to genetic and environmental factors could be calculated. Whereas 39 percent of the variance in reported menstrual flow was accounted for by genetic factors, the corresponding figure for dysmenorrhoea was 55 percent, and for functional limitation from menstrual symptoms was 77 percent (Treloar et al. 1998).

Doctors and Pain

Medical education has not fully adjusted to the evolved model and as part of the diagnostic process doctors often try to distinguish between 'real' pain, meaning that associated with a visible pathological process, and 'psychogenic' pain, where visible pathology is lacking. Behavioural and affective cues are used to interpret the degree of pain and suffering experienced by an individual, which are highly culturally determined and potentially misleading. In the case of cancer pain, cross cultural studies indicate that while associated behaviour and affect may vary, actual suffering is similar in individuals from different cultures. This insight has led to international efforts to improve the availability of strong opioids to cancer patients (see WHO publications 2003).

In clinical practice there is a tendency for patients with certain conditions to be afforded greater respect and appreciation than others. For example, those with long standing low back pain may be viewed with some suspicion especially if the patient is pursuing compensation for a work related injury. Other types of patient may be characterised as 'gallant sufferers' (Selfe et al. 1998). Treatment goals may be very different: value judgements are made as to whether pain reduction, as measured by VAS or questionnaire, is the treatment goal, or whether the objective should be reduction in functional impairment or return to work (Turk et al. 1994). The subjective nature of pain has been emphasised above, but there is a tendency for doctors to underestimate requirements for pain relief in acute pain, for example after surgery, and to overestimate the degree of pain relief by those undergoing treatment for chronic pain. This tendency is taken into account in experimental designs for testing pain treatment outcomes, where reproducible self-report measures are preferred over the clinician's rating. However, in the routine clinical setting there are continuing problems of imprecise assessment and inadequate treatment even when the full range of treatment options is available.

Women, Doctors and Pain

In contrast to men, women experience pain under physiological conditions during normal reproductive life, especially in the form of dysmenorrhoea and pain during childbirth. Certain painful conditions of the reproductive system are specific to women, such as dyspareunia, pelvic pain and endometriosis. The interrelationships between women's experience of painful conditions of the reproductive system, the sociocultural setting and medical models of physiology and pathology represent a fruitful area for research.

In Western societies the broad sociocultural setting is heavily influenced by Judaeo-Christian teaching. In Genesis the current experience of women is portrayed as an outcome of the Fall, in the following versions:

Genesis 3:16 (Septuagint)
και τη γυναικι ειπεν πληθυνων πληθυνω τας λυπας σου και τον στεναγμον σου εν λυπαις τεξη τεκνα και προς τον ανδρα σου η αποστροφη σου και αυτος σου κυριευσει

λυπας *λιπασ* = sorrow
στεναγμον *στεναγμον* = pain with suffering/distress
αποστροφη *αποστροπηε* = separated/apart from

(Vulgate)
mulieri quoque dixit multiplicabo aerumnas tuas et conceptus tuos in dolore paries filios et sub viri potestate eris et ipse dominabitur tui.

(King James)
Unto the woman he said, I will greatly multiply thy sorrow and thy conception; in sorrow thou shalt bring forth children; and thy desire [shall be] to thy husband, and he shall rule over thee.

(New American Standard)
To the woman He said, 'I will greatly multiply your pain in childbirth, In pain you will bring forth children; Yet your desire will be for your husband, And he will rule over you.'

While the Greek has the sense that pain is the cause of distress and the woman's separation from her husband, the English versions suggest a perverse desire, despite pain and suffering in childbirth, leading to a natural subjugation of women. Pain in childbirth has been the norm globally until obstetric anaesthesia became available in the nineteenth century as a result of both technical developments (chloroform/ether) and the influence of opinion leaders (Queen Victoria). The centenary of the licensing of midwives in Britain is currently being celebrated. One consequence of licensing was that midwives were afforded prescribing rights for women in labour not afforded to nurses, and hence were able to administer pethidine and/or nitrous oxide mixtures to women in labour without reference to a doctor. The last twenty years have seen the extensive proliferation of epidural anaesthesia in maternity hospitals in Western countries, such that around 60 percent of primigravidas (women in their first pregnancy) and 20 percent of multigravidas (women in second and subsequent pregnancies) typically use this form of pain relief in labour. Whereas epidural anaesthesia is truly antinociceptive, it is likely that pethidine mainly acts by inducing dysphoria through the delta subclass of opioid receptors (Olofsson et al. 1996). Caesarean section under epidural anaesthesia is extremely popular and its availability is contributing to the rise in U.K. Caesarean rates towards 30 percent in some units. In parallel to the increasing availability of safe and effective technology for analgesia during labour, there has been a growth in demand for non-medical models of intrapartum care. This has led to increased interest in non-pharmacological methods of pain relief such as massage, hot water and TENS (Transcutaneous Electronic Nerve Stimulation). While TENS has been shown conclusively to be ineffective for labour pain it continues to enjoy popularity. Unfortunately, even though excellent evidence exists for the value of the constant presence of a supportive companion during labour, many women worldwide labour alone, unsupported by midwives or by relatives.

Women's experience of labour pain has been the subject of extensive research, numerous interventions and extensive and heated debate. Some of the discourse has focused on the autonomy and rights of women and criticised the medicalisation of childbirth. As discussed in Chapter 4 with regard to private fertility clinics, the location of maternity services within the state or private sectors has influenced approaches to care: in the U.K., the state has on the one hand contributed to medicalisation by strongly encouraging the concentration of births into large hospitals with a full range of medical services (including epidural analgesia), but on the other hand has allowed midwifery to flourish as an independent profession, by measures such as providing indemnity insurance for clinical practice within the health service. Contrasts in the pattern of provision abound between adjacent countries: autonomous midwives deliver the majority of Dutch women while their Belgian sisters experience highly medicalised care.

Practical issues in experimental work on the experience of childbirth have included the extent to which studies can be controlled for the large number of potential confounding variables, and the method of contemporaneous or recall based pain assessment. A key contribution to the literature is the report of Waldenström and colleagues (Waldenström et al. 1996). The authors reviewed the literature on pain recall, emphasising the inconsistent findings from different studies. They highlight an intriguing theme in the literature, the observation that women's reports of pain relief show only moderate associations with the use of different modalities of analgesia. This may be a result of self-selection: for example, women experiencing more severe pain during labour might request more effective analgesia; it is possible that analgesics are used by women to bring pain down to similar levels under different conditions.

In their own investigation of 278 women, 28 percent experienced pain in a positive way, reflecting the generally positive outcome of labour. The descriptors used in pain questionnaire instruments may need to be adapted to include words such as 'happy' as well as the negative terms usually used. Drawing on an early 'Mastery model' (Humenick 1981) for satisfaction during childbirth, the authors reviewed the changes that have been identified in women's preferences before, during and after labour, and the wide differences between individual women in their perceptions and preferences. It was concluded that these differences must be taken into account by birth attendants, who need to decide whether to encourage or discourage the use of analgesia on an individual basis.

Traditional clinical teaching divides dysmenorrhea into two subcategories: primary or spasmodic dysmenorrhea is diagnosed where symptoms have been present since the menarche, where pain starts on the first day of menstruation, and where identifiable pathology is absent. This represents the other physiological form of pain experience unique to women. Secondary or congestive dysmenorrhea occurs at a later stage in menstrual life, pain may build up before the onset of menstruation, associated menstrual disturbance is common, and there is a greater likelihood of specific pathology such as endometriosis. These clinical categories have been used in experimental studies (Amodei et al. 1987) but there may be considerable overlap in practice between the groups. For example, in very young women found to have endometriosis, dysmenorrhea was the most prominent symptom but it was not necessarily of the congestive type (Punnonen and Nikkanen 1980).

Rating of dysmenorrhea in research studies has been accomplished by the use of questionnaires rating a range of associated symptoms as well as pain (Moos 1968), or by the use of pain relief measures specifically for drug efficacy studies (Zhang and Li-Wan 1998). The former type of measure has the advantage of psychometric reliability. It is likely that to some extent the pain of primary dysmenorrhea has a positive element, as discussed above with regard to labour pain. Hypotheses for such positive elements could be conceived in terms of a pain mastery model, or as part of the sociocultural value placed on menstruation as symbolic of female reproductive potency. These aspects have surprisingly not been explored in the social science literature, although the positive side of women's experience of the perimenstrual phase has at least been recognised (Logue and Moos 1988).

Chronic pelvic pain (CPP) affects many women in the reproductive years, with a prevalence of around 15 percent in a U.S. telephone survey

(Mathias et al. 1996). Recent U.K.-based surveys indicate GP consultation rates for CPP somewhat greater than for migraine (Zondervan et al. 1999). Endometriosis represents an important specific pathology contributing to CPP, but many women are left without a diagnosis following laparoscopy and/or other investigations. Systematic study of clinical outcomes identified an effect of doctor–patient interaction as well as 'objective' disease factors such as the presence of endometriosis (Selfe et al. 1998). Women frequently report unsatisfactory and dismissive attitudes during consultations for pain (Grace 1995) and this has been ascribed to inappropriately persistent mechanistic views of pain causation as discussed above (Grace 1998).

Study of attitudinal constructs among gynaecologists identified potential correlates among five constructs with age, gender and ethnicity (Selfe et al. 1998). The construct showing most variation in relation to the study variables was that of 'sociocultural liberalism'. The important unanswered question is whether and to what extent doctors' attitudinal characteristics spill over into consulting behaviour, and whether certain attitudes make for good or bad patient outcomes. As with other types of human interaction it may be that the appropriate matching of doctor and patient is the key to a successful therapeutic relationship: there is no clear evidence that 'liberal' doctors, with consulting styles including a great deal of explanation and choice offered to the patient in treatment decisions, have better treatment outcomes that more traditional colleagues with 'benignly paternalistic' approaches. Certain patients may find the offer of empowerment and choice to be threatening, and would prefer the doctor to make key choices on their behalf. Again, the sociocultural setting is also important: in some cultures it would be considered unacceptable for the doctor to ask the patient her opinion or express uncertainty about diagnosis or treatment. While most British patients have probably become accustomed to their doctor opening a prescribing reference manual before writing a prescription, in many parts of the world this would be regarded as a sign of ignorance.

In Chapter 2 the desires of Sri Lankan clinicians (and politicians) to discourage family planning and abortion because of anxiety about demographic change among the dominant cultural group were discussed. Can any such undercurrents be identified in the context of childbirth and gynaecological care for women with painful conditions? Certainly any residual view that painful childbirth is advantageous seems to have disappeared among obstetricians. However, there are widely differing and presumably culturally determined views among obstetricians about the desireability of Caesarean section for themselves (in the absence of medical need): this was the preference of only 2 percent of Norwegian obstetricians, one fifth of their British counterparts and nearly half of North Americans (Backe et al. 2002).

Another echo of cultural influence can be identified in connection with clinical advice about the timing of childbearing: it is likely that women experiencing severe dysmenorrhoea will experience some amelioration of symptoms after childbearing; similarly, there may be improvement in symptomatic endometriosis. Also, young women with endometriosis may be best advised to start their families earlier rather than later in reproductive life, to avoid the decline in natural fertility associated with age. It may be more difficult for doctors to give and for women to accept such advice in the face of strong social pressures and cultural norms towards later childbearing.

Conclusion

Neuroscience now presents us with new interpretations of pain and its underlying mechanisms, and points us towards an integrative model which fully acknowledges the range of factors involved. Some of the strains in the translation of new understanding into women's health care for physiologically painful circumstances (labour, dysmenorrhoea) and abnormal painful conditions such as chronic pelvic pain have been discussed. The overwhelming need is for primary research into the biological, psychological and cultural origins and meaning of pain in the reproductive organs, whose findings can inform the assumptions and attitudes with which both the patient and her doctor enter the consulting room.

References

Amodei, N., Nelson, R. O., Jarret, R. B. and Sigmin, S. 1987. 'Psychological treatments of dysmenorrhea: differential effectiveness for spasmodics and congestives'. *Journal of Behavior Therapy and Experimental Psychiatry*, 18, 95–103

Backe, B., Salvesen, K. Å. and Sviggum, O. 2002. 'Norwegian obstetricians prefer vaginal route of delivery'. *Lancet*, 359, 629

Berkley, K. and Hubscher, C. H. 1995. 'Are there separate central nervous system pathways for touch and pain?' *Nature Med*, 1, 766–73

Fillingim R. B., Maixner W., Girdler S. S., Light, K. C., Harris, B., Sheps, D. S. and Mason, G. A. 1997. 'Ischemic but not thermal pain sensitivity varies across the menstrual cycle'. *Psychosom Med*, 1997: 59, 512–20

Grace, V. M. 1995. 'Problems of communication, diagnosis, and treatment experienced by women using the New Zealand health services for chronic pelvic pain: a quantitative analysis'. *Health Care Women Inter*, 16: 6, 521–35

—— 1998. 'Mind/body dualism in medicine: The case of chronic pelvic pain without organic pathology – A critical review of the literature'. *International Journal of Health Services*, 28, 127–51

Humenick, S. S. 1981. 'Mastery – the key to childbirth satisfaction – a review'. *Birth And The Family Journal*, 8, 79–83

IASP Task Force on Taxonomy. 1994. 'Classification of Chronic Pain'. H. Merskey and N. Bogduk, eds. 2 ed. Seattle

Logue, C. M. and Moos, R. H. 1988. 'Positive perimenstrual changes – toward a new perspective on the menstrual-cycle'. *Journal of Psychosomatic Research*, 32, 31–40

Mathias, S. D., Kuppermann, M., Liberman, R. F., Lipschutz, R. C. and Steege, J. F. 1996. 'Chronic pelvic pain: prevalence, health-related quality of life, and economic correlates'. *Obstet Gynecol*, 87, 321–7

Melzack, R. and Wall, P. D. 1965. 'Pain Mechanisms: A New Theory'. *Science*, 150, 971–8

Mogil, J. S., Wilson, S. G., Bon, K., Lee, S. E., Chung, K., Raber, P., Pieper, J. O., Hain, H. S., Belknap, J. K., Hubert. L., Elmer, G. I., Chung, J. M. and Devor, M. 1999. 'Heritability of nociception II. Types of nociception revealed by genetic correlation analysis'. *Pain*, 80, 83–93

Moos, R. H. 1968. 'The development of a menstrual distress questionnaire'. *Psychosom Med*, 30, 853–67

Olofsson, C., Ekblom, A., Ekman Ordeberg, G., Hjelm, A. and Irestedt, L. 1996. 'Lack of analgesic effect of systemically administered morphine or pethidine on labour'. *Br J Obstet Gynaecol*, 103, 968–72

Punnonen, R. H. and Nikkanen, V. P. 1980. 'Endometriosis in young women'. *Infertility*, 3, 1–10

Riley, J. L., Robinson, M. E., Wise, E. A. and Price, D. D. 1999. 'A meta-analytic review of pain perception across the menstrual cycle'. *Pain*, 81, 225–35

Selfe, S. A., Matthews, Z. and Stones, R. W. 1998. 'Factors influencing outcome in consultations for chronic pelvic pain'. *Journal of Women's Health*, 7, 1041–8

—— van Vugt, M. and Stones, R. W. 1998. 'Chronic gynaecological pain: an exploration of medical attitudes'. *Pain*, 77, 215–25

Treloar, S. A., Martin, N. G. and Heath, A. C. 1998. 'Longitudinal genetic analysis of menstrual flow, pain, and limitation in a sample of Australian twins'. *Behavior Genetics*, 28, 107–16

Turk, D. C., Brody, M. C. and Okifuji, E. A. 1994. 'Physicians attitudes and practices regarding the long-term prescribing of opioids for noncancer pain'. *Pain*, 59, 201–8

Waldenström, U., Bergman, V. and Vasell, G. 1996. 'The complexity of labor pain: experiences of 278 women'. *Journal of Psychosomatic Obstetrics and Gynecology*, 17: 4, 215–28

WHO publications at <http://www.who.int/cancer/publications/en/> accessed 19th Oct. 2003

Zhang, W. Y. and Li-Wan, P. A. 1998. 'Efficacy of minor analgesics in primary dysmenorrhoea: a systematic review'. *Br J Obstet Gynaecol*, 105, 780–9

Zondervan, K. T., Yudkin, P. L., Vessey, M. P., Dawes, M. G., Barlow, D. H. and Kennedy, S. H. 1999. 'Prevalence and incidence of chronic pelvic pain in primary care: evidence from a national general practice database'. *Br J Obstet Gynaecol*, 106, 1149–55

CHAPTER 6

LABOUR, PRIVATISATION AND CLASS:
MIDDLE-CLASS WOMEN'S EXPERIENCE OF
CHANGING HOSPITAL BIRTHS IN CALCUTTA

Henrike Donner

Introduction

Hospital births were introduced to India during the colonial period but only became popular among the Calcutta middle class during the 1960s. Today they have replaced the 'traditional' home birth among the more affluent urbanites, and more recently a significant proportion of hospital births have taken place in private institutions. The introduction of private healthcare providers in the wake of liberalisation policies is commonly held responsible for changes to the way reproductive health is understood and, in particular, the ways contraception and deliveries are handled. Here, as in many other contexts, a high number of 'elective' Caesarean sections paralleled the proliferation of private healthcare and it has been suggested that changes towards interventionist medical procedures result from the development of these new markets.[1]

Following earlier explorations of 'other' models of childbirth, scholars have recently focused on the effects the imposition of 'medicalised childbirth' has on existing practices and meanings in a wide range of regional settings. This interest in the exploration of non-traditional, medicalised childbirth (see Sesia 1997, Szurek 1997, Hunt 1999) is fuelled by a focus on the 'cultures of globalisation' and the diversity of effects the adaptation of 'Western' practices and knowledge may have. It is apparent that a more differentiated view on technologies and the so-called 'new reproductive world order' has emerged, and although a wide range of studies emphasise the negative consequences of reproductive change, authors like Ginsburg and Rapp argue that:

> Reproduction also provides a terrain for imagining new cultural futures and transformations through personal struggle, generational mobility, social movements, and the contested claims of powerful religious and political ideologies. These imaginings and actions are often the subject of conflict, for they engage the deepest aspirations and the sense of survival of groups divided by differences in generation, ethnicity, race, nationality, class, and, of course, gender. (Ginsburg and Rapp 1995: 2)

This more encompassing approach has already provided new insights where 'new reproductive technologies' are concerned (see Edwards et al. 1993, Inhorn 2002), but it has still to be employed in the study of 'normal' birth itself.

This chapter deals with hospital births in Calcutta and focuses on the minority of middle-class women with privileged access to healthcare services. The provision of these services has expanded over the last twenty years, to a significant degree, which has inevitably affected birthing among those who routinely use these facilities.

According to a longstanding argument, medicalised models of reproduction and interventionist technologies are particularly successful where a market, such as the one that emerged in the Indian context in the aftermath of economic liberalisation from the early 1990s onwards, developed. Here as elsewhere, the deficits of state-run healthcare services led to a proliferation of private sector undertakings, which mostly, but not exclusively, cater to the needs of the ever-growing, affluent middle class. According to websites like www.tradepartners.gov.uk, aiming at investors, the healthcare market in India is expected to grow at an annual rate of 13 percent until 2005, enough to attract investment from national and international sources to an industry worth £10.4 billion. Much of this investment went into hospitals and nursing homes as well as specialist clinics dealing with reproductive medicine, an area where interventionist technologies are now routinely used in urban areas as well as among the more affluent strata of rural populations (see International Institute for Population Sciences (IIPS) and ORC Macro 2002).

In this chapter I focus on the domestic and kin relationships that shape women's access, use and evaluation of modern healthcare as part of class-based identities. Such an investigation of a 'domestic perspective' and the way healthcare is influenced by relations within the household is relevant if we are to analyse new technologies and changing reproductive patterns in relation to the wider implications economic change and state policies have in the lives of women.

Many analysts of the recently emerged markets in private healthcare explain changes in values and consumption patterns with reference to the increased status competition and desire for upward mobility that allegedly dominate middle-class life in urban India (Dwyer 2000, Fernandes 2002). Instead, I suggest that an analysis of these patterns must situate different birthing experiences and the medicalisation of childbirth within a specific domestic setting that affects women's experiences and ultimately determines their sense of agency. What is investigated in this specific context is why women choose interventionist deliveries, given that they 'collaborate (...) because of their own needs and motives, which in turn grow out of the class-specific nature of the their subordination' (Riessman cited in Jacobson 2001: 222–23).

Setting the Scene

The maternal histories on which this chapter is based refer to middle-class women, although during some of the interviews working-class women, for instance maidservants, were present and contributed to the discussion. The approach adopted here features some of the characteristics detailed by Riessmann, who emphasises the reliance on interview excerpts, the attention to the structural features of discourse and a comparative approach to interpreting similarities and differences between participants (Riessmann 2002: 152–53). Except for medical surveys, such as the study of hospital births in Chennai undertaken by Pai, Sundaram, Radhakrishnan, Thomas and Muliyil (see Pai et al. 1999), such material is rare and the available studies lack the detailed contextual analysis that long-term fieldwork can provide.[2]

A focus on the middle classes is relevant because of this site of the middle classes, which also serve as a reference group for marginalised communities and 'the poor'. Perceptions and attitudes of middle-class men and women who are involved in the promotion of concepts and services as consumers, medical professionals, planners, and social workers, are crucial for the development and implementation of government programmes and private enterprise in the health sector through which choices are made available to those who are less privileged. These trickle-down effects are particularly pronounced in urban areas, where the lives of the expanding middle classes are closely linked with the lives of the urban poor (Dickey 2000).

In this context the term 'middle class' is based on the self-definition of those interviewed, who ascribe to common denominators of status. Although the sample is heterogeneous in terms of economic standing, all women interviewed agreed that a common 'culture' of domesticity which emerged in the colonial period is characteristic for the middle class, especially among members of the Bengali-speaking community.[3] Like Dickey's more affluent informants in Madurai, they referred to a tripartite model consisting of the poor, the middle class, and the upper class (Dickey 2000: 465). Thus, the coherence of the sample derives from shared values, consumption, marital, occupational and residential patterns, which overrule economic divisions, variations of caste and often even the preference for endogamous marriages (Béteille 1996, Donner 1999).

The argument is based on interviews with women in thirty-seven households during extensive periods of fieldwork between April 1995 and

August 2000, and the material was collected in the course of semi-structured interviews in two neighbourhoods of Calcutta. Most of the women interviewed in the course of the fieldwork were Bengali-speaking Hindus, a sizeable minority were Bengali-speaking Christians or from north-Indian business communities.[4]

One neighbourhood is a mixed and densely populated area (Nair 1990: 13ff, Mukherjee 1993: 22–48), whereas the second emerged after the partition of India with the settlement of East Bengali Hindu refugees, and is therefore more homogenous in terms of ethnic origin and class (Bose 1968: 33–4). All but two women presented their stories in Bengali, and often the material I refer to as 'maternal histories' formed part of wider narratives.

During fieldwork, only four births occurred in these households, partly because many women were already mothers of adolescents, and partly because the birth rate in middle class households is generally low (none of the women who gave birth during the past ten years has had more than one child). In two cases I was able to visit the pregnant woman/young mother before and/or after the delivery, and visits in hospital and further informal conversations with friends and medical practitioners provided ample evidence to support the generalisations drawn from the interviews. However, it is important to bear in mind that the focus of the interviews was on their experiences and the context within which they and their relatives placed childbirth, rather than the pros and cons of specific procedures and medical knowledge.

Most of the households included members of two or more generations, and this extended or 'joint' setting is the stated ideal of all middle-aged women. Younger, recently married women hope to form a nuclear household at one point, but the patrilocal norm ensures that most women spend a considerable time, if not their whole lives, in their parents-in-law's house. Not surprisingly, women's emphasis is clearly on 'traditional' family values, and in this pronatalist setting the orientation towards the roles as mothers and housewives is very pronounced (Chatterjee 1993: 35, Standing 1991: 63ff, Varma 1998: 166). It is generally assumed that marriage and parenthood provide meaningful lives, and while a career may be an alternative for unmarried women, the values attached to parenthood are part and parcel of the socialisation of boys as well as girls. Although some of the younger women were working 'outside', the percentage of married middle-class women in employment is very low in Calcutta, and the majority will leave employment on the birth of their first child.[5]

The following case studies present the experiences of two women belonging to different generations, who gave birth in hospital, and highlight changes in the organisation of hospital deliveries. The first case study relates to deliveries in what I call the early phase of hospital births, starting around the middle of the 1950s, and the material is compiled from different interviews during which these experiences were recollected. The second case study is the more detailed account of the circumstances surrounding the birth of a baby with whose paternal family I have been working since 1995. This delivery exemplifies the common trend towards deliveries in a large private institution and the immense popularity of 'elective' Caesarean sections.

Case Study I: Ila Chakraborty

At the time of the interview, Ila Chakraborty was in her mid-fifties and lived with her husband and two sons in an affluent South Calcutta suburb. Like most of their neighbours her family were refugees from East Bengal. She married at the age of nineteen and spent many years with her parents-in-law before her husband, who is a contractor, could afford to build the very spacious house they inhabit today. During that earlier period in her parents-in-law's house she had her three children, and Ila emphasised that her relationship with her mother-in-law and elder sister-in-law was good for many years. She presented her maternal history in the light of her 28-year-old daughter's marriage and the birth of her granddaughter, which were very important themes in her life at the time of the interviews. The marriage of her daughter had been arranged with the son of a business friend of her husband's, who lives with his parents, brothers and their families in a neighbourhood close by.

When Ila herself married, she was the younger of two daughters-in-law in a household consisting of three married couples, their children and unmarried relatives. One of the complaints about this time, which she sees as generally happy and peaceful, was that her mother-in-law rarely allowed her to visit her parents, who lived a short bus ride away. However, as her parents-in-law and her parents got on very well, she received visits from her father, and she was keen to point out that her own daughter's in-laws make her feel equally welcome. In an established pattern, she or her husband visit the daughter's new home daily, because she is often preoccupied with housework. This does of course repeat the pattern Ila herself experienced in her parents-in-law's house, where, as she stated, the daughters-in-law were responsible for the children and were rarely allowed to visit their paternal home. According to her, the way work was allocated in the house also explains, why she was not allowed to return to her parents' house for the birth of her first child, as used to be customary:

> When I was pregnant I did not go to my natal home, but remained in my in-laws' house, although this is our custom and women should not work for a month after the birth. I gave birth to my daughter in a nursing home nearby and the sons were born in another one. I did my two check-ups during each of the pregnancies but everything was fine, I did not have any problems – not like my daughter who needed months of bed rest. Of course I suffered from nausea, but that did not stop me from working, because my mother-in-law did not allow me to work less – so I worked until the end.

Reflecting on her deliveries she said:

> When the pain came, I told my sister-in-law, and when it got worse and worse I was brought to the nursing home by my brother-in-law. At that time husbands, or their parents, would not accompany a woman about to give birth. My mother-in-law was too old anyway, she came the next day to look at the grandchild. There was a nurse with me, who did all the check-ups, but the doctor delivered the child. I had spent a lot of time worrying, what would happen, whether I would be able to cope with the pain, whether the child would be all right ... at that time we didn't know whether the child would be a boy or a girl – we didn't have ultrasonography then.
>
> Today doctors can tell, but they won't say – in my daughter's case the doctor could tell – he saw it and said 'look here is the head, here is a leg', and her husband said to her that he believed it would be a girl. They used to tell you the sex, but then there were these arguments about it and since then they will not tell anymore.[6] Because some abort girls they stopped telling you ... When I gave birth to my daughter I was already in labour when I arrived at the nursing home, that was around nine o'clock in the evening. They said then after some hours that they would use forceps to get her out, because it took so long. It was terrible and very hard. In the case of my first son the waters broke, and I went to the nursing home, but nothing happened so they gave me an injection because I had no contractions for two days after that. They thought they would have to do a '*Caesar*', but that did not happen and after three hours he was born.

Like others, Ila then described the birth of her second son in detail and concluded that she had enough after this delivery, 'I thought I won't do that again – never again'. She remained in hospital for two days after each delivery and had a check-up some time later, but since there were no complications this seemed of little significance to her. When she returned to her parents-in-law's house she was allowed to 'rest' for about two weeks. When the customary period of confinement (and pollution) was officially ended by a Shasthi *puja* on behalf of the newborn, she had already returned to the hearth and resumed her duties.

Case Study II: Madhushree Chowdhury

Madhushree, who gave birth while I was doing fieldwork in 2000, at the time lived next door to the Chakraborty family in her parents-in-law's house. She had been married for two years and was three months pregnant when I began my fieldwork in October 1999.

At the time, the Chowdhury household consisted of Madhushree and her husband Amrit and his parents, as well as a live-in servant. Madhushree was in employment – working in the billing department of Nokia telecommunications when she became pregnant, while her husband worked as a representative for another multinational company. His mother had been the manager of a small but prestigious nursing home before her children started secondary school, his father is a government engineer.

When I decided to visit the Chowdhury family on the 25 February around noon, I found Mr Chowdhury fast asleep and Mrs Chowdhury had just taken her morning bath and attended to the deities. Upon seeing me, she exclaimed that I had chosen a good and auspicious day to come, since Madhushree had been delivered of a boy in the morning. We settled down for a cup of tea in the living room, and she began to recount the events of the last twenty-four hours, leading up to the birth of her first grandchild.

Madhushree had been in her mother's house for some weeks, and had called the previous evening and complained about discharge, so Mrs Chowdhury contacted the doctor who offered to perform a Caesarean section early the following morning, as her medical 'team', consisting of this female obstetrician, an anaesthetist and two nurses, would be assembled in one of Calcutta's most prestigious nursing homes to perform a series of Caesareans. Working from midnight onwards, they could easily 'slot' Madhushree in. After many telephone calls Madhushree, her mother-in-law and her father-in-law had agreed to proceed, though regrettably Amrit, the father-to-be, was on a training course in Mumbai.

Around 5.30am, the Chowdhurys had gone to Woodlands Nursing Home, where they met Madhushree, her brother and her mother. After assisting her with the formalities, the visitors settled in her room, and shortly afterwards she was wheeled into the operating theatre. About twenty minutes later, the relatives were joined by Amrit's sister, her husband and a female cousin, when the doctor came and announced the birth of a baby boy, his weight and the exact time of birth. Then the newborn was brought in by a nurse, who ensured that a male elder checked the sex of the baby before he was transferred to the adjacent baby room.

Mrs Chowdhury's account ended here and she got up to make phone calls to relatives but invited me to join them for the afternoon trip to the hospital.

The nursing home, located in one of the best areas of Calcutta, is housed in a purpose-built building, which reflects the tastes of its affluent private clients. From a huge marbled hallway, elevators and a staircase lead to the wards, which are modern, well-maintained and consist of spacious rooms with up to four beds. When we entered her room, Madhushree was awake, and on seeing her mother-in-law complained about thirst and pain in her lower abdomen.

Shortly afterwards Madhushree's mother and her brother arrived and were led in by Mr Chowdhury. One hour later (and ten hours after the 'operation') the room was crowded with relatives, friends and neighbours, who had come to look at the newborn and meet his mother.

After an interval of three days I returned to the hospital with my partner and 1-year-old son, where we met the Chowdhurys, Madhushree's mother and brother and some friends, who were chatting with Madhushree. She was still propped up against some pillows but said that she felt much better. Her main complaint was that the nurses did not bring the baby in often enough, as 'rooming in' is not offered by exclusive hospitals like this, but she obviously enjoyed the attention of her guests immensely.[7]

However, among the affines the sense of relief, which had united her parents-in-law and her mother and brother during my first visit, had given way to a palpable tension. This was caused by the question of where Madhushree and (in the eyes of her parents-in-law more importantly) the newborn would go once she was released from hospital. Though the matter was not discussed openly, we witnessed a silent battle and got to know that both camps had stocked up on equipment and extra help to take care of mother and child.

In the end, the Chowdhurys were overruled by their strong-willed daughter-in-law, who took her son to her mother's flat arguing that custom allows a young mother to spend some time in her natal home before and after the birth. Her parents-in-law were forced to look on as Madhushree manipulated notions of customary rights and filial duty to her advantage, and even worse, had to accept that Amrit also spent the month-long 'confinement' in his wife's mother's house. The conflict resulted in an argument between father and son on the occasion of the *puja*, during which the blessings of the goddess Shashti are sought on behalf of the newborn. This event was organised in the Chowdhurys' house, and involved relatives, friends and neighbours, who assembled to greet the young couple and baby Akash, who was finally welcomed into his paternal patriline (*bangsha*) and arrived at his future home.

Modern Reproductive Regimes

The changes reproduction has undergone in 'modern times' result from institutional changes as well as new medical procedures and technologies. 'Native' practices relating to kinship, marriage and the raising of children were scrutinised by foreign administrators, missionaries and social reformers during the colonial period (see Chatterjee 1993). In many instances European concepts of pollution, wellbeing and household organisation, were imposed upon women whose maternal qualities were questioned on the basis of racist stereotypes (Jolly 1998: 6). However, although hegemonic discourses on 'hygiene', 'antenatal' and 'postpartum' care were influential and inform policies of postcolonial states to date, the diverse range of reactions – from direct resistance and indirect protest to the partial appropriation and transformation of 'Western models'– is well documented throughout the region (see Ram and Jolly 1998).[8] These studies as well as the chapters by Unnithan-Kumar and Simpson in this volume demonstrate that class is an important variable in the analysis of such processes.

In the case of urban Bengal, middle-class notions of the role of 'modern woman' emerged from the middle of the nineteenth century onwards with the redefined boundaries between the public and private sphere, which where policed by male representatives of different communities.

But until hospital deliveries became more widespread, pregnancy and birth did not figure as prominently in these discourses as female education, conjugality and progressive motherhood, although some medical practitioners and reformers raised the issue of high maternal mortality rates among secluded upper-caste women (Borthwick 1984, Chatterjee (1992) 1995, Walsh 1995). Even though middle-class women gradually entered the public sphere as pupils and students, political activists and professionals, they remained literally hidden from public view and intervention as birthing women during the colonial period and beyond.[9]

The institutionalisation of hospital births on the Indian subcontinent commenced more than a hundred years ago, when the first maternity wards were established in Madras and Calcutta during the later half of the nineteenth century. However, the influence of these medical institutions remained limited, due to the influence of the upper-caste norms of chastity and seclusion. Anecdotal evidence, which due to the lack of quantitative data is the only source available, strongly suggests that routine antenatal services and hospital deliveries among the middle classes in metropolitan areas did not emerge until the 1960s (Engels 1996). Since then, hospitalised birth has become the norm, and in line with comparable developments elsewhere, 'elective' Caesarean sections have become the delivery of choice: 'The proportion of deliveries by Caesarian section was three times as high in urban areas (15 percent) as in rural areas (5 percent). Among births delivered by health professionals, 20 percent in urban areas and 15 percent in rural areas were delivered by caesarian section' (International Institute for Population Sciences (IIPS) and ORC Macro 2002: 299).

Furthermore, the same report published by the Institute for Population Sciences states that the percentage of such deliveries increased substantially between the two surveys, namely from three to seven percent in the period between 1992–1999. This process, which mirrors earlier developments in comparable contexts (Belizán et al. 1999, see Pai et al. 1999), clearly demands a closer look, if one is to understand what 'medicalisation of childbirth' may imply in this context.

Private Healthcare

The described pattern of hospitalised childbirth underwent further dramatic changes through the introduction of new technologies and the recent growth of private healthcare in urban areas. Contrary to a wellworn argument, the use of private healthcare was widespread before economic liberalisation took off. But since then laboratories and nursing homes mushroomed all over India and multinational companies as well as small scale entrepreneurs invest in this booming and largely unregulated new sector. These institutions are in many instances located in affluent neighbourhoods or the spreading suburbs and range from small units with just a dozen beds to large developments on brown-field sites developed by companies like Apollo Group, Wockhardt and Duncan. The latter feature prominently in the imagination and local lore, as they represent exclusive modernist spaces, an image reflected in their design and organisation.

The relationship between the middle classes and this largely unregulated private healthcare system is rather ambivalent. While the status attached to the use of specific reputed institutions and the services rendered is high – rarely would a middle class family choose a government hospital – the accompanying profit orientation and alleged corruption fuel distrust and contempt.

Public opinion of those providing healthcare, including doctors, nurses and the management of nursing homes, is very low and conflicts between staff and patients testify to the tensions arising from the perceived lack of credibility – enhanced by the unquestioning support unions extend to employees. In this rather problematic environment personal relationships between doctors and patients are vital, and hold the promise of sufficient services and care.[10] Far from being a mere consideration of financial means, doctors and nursing homes are chosen because the family of the potential patient knows the doctor working there, and feels more secure in the knowledge that he or she takes a personal interest. This produces a double standard in public discourses, where private healthcare is concerned, and while doctors are accused of greed, individual households cultivate personal links with specific doctors, who are seen as reliable and whose advice, even if it includes costly treatments and tests, goes unchallenged.

All women interviewed except for one pointed out that the doctors who delivered their babies had been personal acquaintances of their husbands' families or their parents and that they provided them with the best of services. The only instance where doubts regarding the performance of doctors were raised was the case of a woman who was maltreated by her parents-in-laws and lost her first baby but was still forced to go through 'normal' birth. Given the preference for private healthcare it is hardly surprising that antenatal care and deliveries have become more elaborate in the wake of the development of markets based on the interplay between doctors, laboratories and hospitals. Among the most prominent changes women registered are the intensification of antenatal care and testing during pregnancy and the trend towards 'elective' Caesarean sections.[11]

The link between reproductive technologies and the privatisation of healthcare has been seen as the main explanatory model for the spread of interventionist procedures, in particular elective Caesareans in developing societies (see Potter et al. 2001). Furthermore, since the conjunction of state and privately provided healthcare with the low status of women in many South Asian contexts is seen as a reason for the routine employment of amniocentesis, abortion, intra-uterine devices and hysterectomies (Patel 1989, Unnithan-Kumar 2002), such explanations seem to fit feminist analysis of new reproductive regimes. The underlying 'production view' of the female body linked to the imagery of the body as a machine has been identified by feminists like Martin as linked to the acceptance of Western obstetrics (Stephens 1986, Martin 1987). But the influence of established allopathic notions notwithstanding, mechanistic metaphors and functional approaches are not the preserve of 'Western perspectives'. The medicalised view on women's reproductive bodies fits surprisingly well with the ayurvedic understanding of the body and its functions as a system of channels and substances which can become unbalanced/ blocked. While the interviewees' understandings of the processes underlying fertility and conception is very limited, they have a very detailed knowledge of complex technologies and medical procedures used to deal with reproductive problems, including deliveries. In this, as in other South Asian settings, amniocentesis, IUDs and Caesarean sections are interpreted as blessings of modernity which enable women, and increasingly couples, to enhance their chances of having the desired child of the right sex at the most suitable time (see Unnithan-Kumar in this volume and Handwerker 2002). In the worldview of those interviewed, pregnancies and birth have been brought under the control of medical professionals, who help women to achieve these aims and fulfill their desire to become the healthy mother of a healthy son.

Dominant Themes: Childbirth Past and Present

Pregnancy and childbirth are complex processes, but individual representations like maternal histories are dominated by a number of interrelated themes. One of them is the opposition between tradition and modernity, an idiom used by women when describing the changes initiated by the shift from home birth to hospital birth. For them, as for the less well educated, changing childbirth practices are part of the complex imagery of modernity, alongside transformations of conjugality, the education of girls, and women's political participation (see Borthwick 1984, Chatterjee 1993, Engels 1996). Thus, individual time, chronological time and notions of development are often described with reference to changing reproductive practices, and in the course of many conversations the availability of new reproductive technologies signified intergenerational differences.

But while the dichotomy between past and present may appear clearcut in hindsight, there are strong continuities as well, as actual practices were modified within a set of structural constraints – most prominently the norms associated with patrilocality. Maternal histories testify to this as most older women begin their recollections with accounts of the acute sense of shame felt by a young married woman once she realised that she was pregnant. The early part of pregnancy was spent in the parents-in-law's house and was characterised by food taboos and restrictions regarding the mobility of the pregnant daughter-in-law. Among secluded upper-caste women, where contact with strangers and male relations was restricted at any time, it became even more minimal during these periods. In addition, precautions against dangers stemming from ghosts and the evil eye, as well as contaminated food and unhealthy thoughts, had to be taken. Whereas supernatural threats have become less prominent and young pregnant wives can go out at night without fear of ghosts, restrictions on the variety of food consumed, movements and exposure to unrelated persons, are still imposed by many a mother-in-law.

The issue of control over one's own pregnant body and delivery was prominent in the past, when a young wife expected her mother-in-law to lower her workload, provide her with healthy food, and organise rituals promoting the formation and growth of a male foetus and an easy delivery (*shadh*).[12] Like other north Indian communities, Bengali women were traditionally sent back to their parents' house a couple of weeks before the birth of at least the first child and remained there until the postpartum period of confinement was over. This practice has been on the decline in middle class families, because, as many senior women observed, the right of the affines to refuse their daughter-in-law visits to her parental home has become legitimised by the demands of continuous antenatal care. Since the 1960s doctors are involved from the earliest stages of pregnancy onwards, and consequently they are chosen and paid for by the parents-in-law. Consequently, a young, married woman will remain in their house until she goes to hospital and return there afterwards. In the process, a very important positive aspect of women's experience of childbearing – namely the role played by natal kin, her return to her parental home to give birth, and the subsequent period of rest surrounded by 'one's own people' – has declined in the urban areas (Raheja and Gold 1994: 73ff).

Moreover, in the period before nursing homes became popular, birthing was an all-female affair. At the onset of contractions, referred to as *baeta* (pain), a pregnant woman went into a separate room, which had been equipped with a pile of rugs or sometimes an old bedstead. It was up to the birthing woman's senior relatives to decide when to call specialists – but they normally sent for a *dai* (traditional birth attendant), though according to some this often happened quite late. Sometimes a senior kinswoman remained in the same room with the birthing woman throughout the delivery, but on most occasions fear of pollution prevented any involvement of relatives. Given the fact that a *dai* was in all these families regarded as a low-caste working-class person called to deal with a polluting and potentially dangerous event, her effective support and autonomy in dealing with the birthing woman was very limited (Rozario 1998).[13]

Since the advent of hospital birth women were booked into a state-run or private nursing home early during the pregnancy and were expected to arrive there when the contractions set in. During the early phase of hospital birth, such institutions were mainly chosen within the same locality, unless complications were expected and a pregnant woman was seen by a specialist in a large hospital.[14] As hospital births demanded contact with the outside sphere, male intermediaries were increasingly involved in the organisation, and birthing women arrived at the nursing home accompanied by senior male affines.

For the birthing women this generally implied that the shame about pollution – enhanced in the often hostile environment of the parents-in-law's house – was substituted by the shame caused by the presence of a (normally avoided) husband's elder brother, father or uncle, who would sit in the corridor during the delivery. This acute embarrassment was certainly shared by the men responsible for the birthing women.

Once admitted, the birthing woman was routinely confined to a 'stretcher' and attended to by a *dai* until the delivery was imminent. According to many accounts the *dai's* positive impact was even more limited here than in the house as the social distance between her and the birthing woman was reinforced by the institutional hierarchy.[15] As a hospital employee, the *dai* ceased to provide moderate measures for pain relief like massage, loosening of the hair and clothes and stuck to cleaning, inserting catheters, cutting and disposing of the umbilical cord.[16]

In most cases a nurse supervised the early period of labour and was responsible for standard procedures like manual external examinations. The appointed doctor arrived once the actual delivery was imminent, but often reached the nursing home only after the birth. Women belonging to this middle-aged generation presented doctors and nurses as powerful, knowledgeable and proactive. But while one of the positive aspects of hospital birth was the successful management of pollution outside the home, the involvement of its employees in the care of mother and child was described as less desirable. In fact, it was among the reasons women cited for leaving the nursing home as soon as possible.

Returning to their parents-in-law's house, women were put up in a room and often an extra servant was employed to look after mother and child for the period of postnatal confinement, which lasted up to thirty days depending on the caste background of the family. But as an increasing number spent this period in the affines' house the most enjoyable aspects of confinement – the rest and food a mother would provide for her daughter under the same circumstances – turned into a matter of control. Under the new regime affines decided what was best, and in more than one case did not provide the expected services at all. Some new mothers demonstrated their dissent and rage over this new pattern by throwing away food cooked especially for them, others refused to eat 'healthy' food altogether. However, none of them could escape considerable isolation, which demonstrates that postnatal confinement, a period of positive adjustment and recovery in other contexts (see Patel 1998), takes on different meanings according to circumstances.

According to the women interviewed, yet another shift occurred with the introduction of elective Caesarean sections, which have become the delivery of choice among younger women.

More so than hospital births in general, Caesarean sections represent a 'Western' and therefore contemporary mode of delivery, and women reacted with surprise and disbelief when told that this type of birth is not the norm in 'the West'. The prestige attached to a modern, hightech delivery has rightly been highlighted by researchers investigating the 'modern medical hegemony' of hospitalised deliveries (see Georges 1997: 94–6). In recent studies the status attached to hightech medical procedures is held responsible for the rising rates of elective Caesareans among middle class women in Latin America (Belizán et al. 1999), Turkey (Tatar et al. 2000) or India (Pai et al. 1999). Since doctors have an interest in performing interventionist and expensive procedures, recent research has highlighted that women in many of these settings are subjected to Caesarean sections even if they may prefer a vaginal delivery (Potter et al. 2001).

But apart from this aspect, the women interviewed attributed the effective management of pain to Caeasarean sections. In comparison, vaginal deliveries are depicted as agonisingly slow and potentially at least as dangerous as Caesarean sections.[17] Unlike Van Hollen's lower-class informants in Chennai, who emphasised the strengthening effects of normal birth, a discourse on the positive aspects of *shakti* (female power) manifest in the strengthening experience of birthing, is not prevalent in the given setting (Van Hollen 1998: 160). Furthermore, although women were well aware of the interest medical practitioners have in a scheduled and speedy delivery, the image of Caesarean sections as less painful and 'safe' in comparison with the so-called normal (*emni*) birth is of great importance. In spite of learned debates regarding the safety and pain experienced during and after different types of delivery, it is widely believed in urban Bengal that women suffer more during 'normal' births.

In the course of the interviews pain was a much discussed issue and mothers who had 'normal' births did not hesitate to describe the excruciating pain, length and problematic nature of birth in general and the deliveries of their own children in considerable detail.[18]

No doubt, such narratives of senior women's experiences fuel a negative attitude to 'normal' births, although some of the older women asserted that giving birth becomes easier with practice. Debates concerned with various types of deliveries often centre around the pain and lasting harm caused by different modes of delivery. While none of the women suggested that Caesarean sections do not cause pain and make women suffer, it is the shame, pain and pollution associated with 'normal' birth young women avoid by having Caesarean sections. By comparison, a fast, scheduled and specialised procedure seems desirable and provides a sense of control, in a setting in which women do not expect birthing to be an enjoyable experience. If we include shame and the resulting psychological suffering into our definition of pain, it is clear why women assert that hospital births allow for the successful management of pain by relocating the pollution of birth outside the house. Whether or not Caesareans are de facto more painful and high-risk – as many commentators assert – is not really the issue. It became clear in the course of the interviews that, rather than looking at the scientific or medical evidence supporting such assumptions, class and intra-household relations provided the framework within which women argued in favour of one or the other type of delivery.

Positive Aspects of Modern Deliveries: Domestic Relations, Pollution and Knowledge

These views of past and present childbirth experiences direct our attention to various types of authority and knowledge involved, which has been the focus of the feminist critique of medicalised childbirth. Since the publication of Brigitte Jordan's seminal comparative work, 'Birth in Four Cultures' (Jordan (1978) 1993), numerous cross-cultural studies of hospital births have stated the alienating effects of the new authoritative knowledge associated with it. Jordan's summary set the tone for the critical work on 'cosmopolitan obstetrics' when she wrote in 1978:

> In many countries of the Third World, strategies for development include the importation of obstetric technology and of technology-dependent obstetric procedures such as hospital deliveries, pharmacologically managed labours, the use of ultrasound and electronic foetal monitoring, induction of labour, instrumental and surgical delivery, and the care of premature and sick infants in intensive care units. While it is clear that such facilities and their technologies will lower some kinds of mortality and morbidity, their importation often has unforeseen and not really assessed negative effects. Beyond that, the replacement of traditional 'low technology' raises fundamental questions about concomitant transformation in the nature of knowledge about the birth process, which in turn affects the distribution of decision-making power and the ability of women to control the reproductive process. (Jordan (1978) 1993: 199–200)

The critical study of medicalised childbirth in poor and marginalised communities in South Asia has all too often depicted a system imposed on poor women, with mostly negative and often deadly consequences. But the continuously high rates of mother and child mortality[19] and detailed studies of reproductive health based on fieldwork in rural areas do not support a nostalgic view of other than allopathic approaches to childbirth in South Asia (Jeffery et al. 1989, Rozario 1998). More and more poor women in rural as well as urban areas have access to antenatal services and hospitals, and use multiple health seeking strategies, including the interventionist reproductive technologies, many of which are employed to produce children of the desired sex, and may put those in poor health at risk (see Van Hollen 1998, Unnithan-Kumar 2002).

In the given context, the majority of younger women see hospital deliveries as a safe and appropriate form of delivery. Apart from pain management they represent to them a way to enhance and negotiate individual rights in the parents-in-law's house. The senior generation in particular interpreted the more recent form of hospital deliveries as a way to make up for the loss of customary rights. It is the authority of medical practitioners, which provides the basis for negotiating their workload during pregnancy and the postpartum period.

Even where, like in the case of Madhushree, young women have access to better doctors and hospitals, not because their parents or parents-in-law grant them better treatment, but because their husband's status entitles them to specific insurance services, medical knowledge as an institutional framework situated outside the domestic domain is held responsible for the advantages attributed to 'modern' deliveries. As Unnithan-Kumar's study of the attitudes of poor women in Jaipur to reproductive technologies has demonstrated, their acceptance is guided by kin relations and household organisation. Thus, these are crucial sites through which women negotiate reproductive change, the use of technologies and their own sense of agency.

The division of work in the household figures prominently in the accounts Jeffery, Jeffery and Lyon collected in rural Uttar Pradesh, where it is the most obvious factor determining the type of medical care birthing women receive among the less affluent (Jeffery et al. 1989). But although the control over a young woman's labour power is as pervasive a theme in the middle-class families we are concerned with, among whom the actual work undertaken fits Papanek's description of status production work (Papanek [1979] 1989), I would like to caution against an oversimplification of such an argument. As Sangari (1999) observed with reference to class and labour 'Though similar in terms of responsibility, the contemporary middle-class woman's assisted labour, often geared to family status production including activities related to high consumption and elaborate maintenance of property, is not comparable to poor rural or urban women's survival-orientated domestic labour, essential to the bare existence of the family.'

More importantly, she asserts, are the relations between women belonging to different classes and the way work is divided between them 'There is, however, a relation between different types and definitions of domestic labour along the joint axes of class differentiation and mobility' (Sangari 1999: 295).

This relational view is brought sharply into focus where different types of deliveries are concerned, and I suggest that this angle provides the basis for a more fruitful exploration of hospital births and the rise in Caesarean sections.

Significantly, all women interviewed, including Madhushree, presented their experiences of pregnancy and birth in the context of their role as daughter-in-law living with affines. These young women, whether they are in employment or not, have multiple responsibilities in their parents-in-law's households, and because they are expected to leave employment after the birth of a child they prioritise this role when the pros and cons of different deliveries are discussed.

In spite of her class position a daughter-in-law may be expected to contribute to the preparation of food, the supervision of servants and the education of children soon after birth (see Donner 1999).

Thus, one distinct advantage of Caesarean sections in the view of middle-class women is that a young mother cannot possibly be expected to resume her duties right after the operation. This point came out even more clearly in discussions with working-class women – for instance those working in the households concerned – who highlighted the negative effect interventionist deliveries have on their labour power. The same physical effects anticipated as a chance to gain extra help, rest and support by middle-class daughters-in-law, are avoided by those whose livelihoods depend on manual labour. This demonstrates that the technology itself or the hegemonic discourse on medical consequences are relevant but not predominant where actual decisions regarding different types of delivery are concerned. Women belonging to different backgrounds negotiate reproductive technologies in the context of prevailing patterns of residence and intra-household hierarchies. Concerns with expenses and status in the narrow sense come second to these more complex considerations of a woman's role in the home, and they evaluate the merit of specific procedures circumstantially.

Conclusion: Home, Kin Relations and the Meaning of Intervention

This chapter provides new material on the way a reproductive regime and the technology that goes with it, are contextualised within a non-Western, but privileged setting. In that the reproductive technologies implied are hightech, questions asked by those studying assisted conception and technologies dealing with infertility apply here as well. However, we still have to ascertain, what happens, when so-called Western models are exported and rather than depicting those who are confronted with them as 'others', we are to analyse them as co-participants 'in the same historical trajectory' (Inhorn 2002: 20).

In this critical study, the discussion and the case studies highlight the importance of intra-household relations, which shape maternal histories and decisions regarding reproductive technologies. These shape the way hospital births and various interventions are used and evaluated and let the notion of the 'home' (employed in the opposition of 'home' and 'hospital') appear less self-explanatory. A closer look at the effects of patrilocality and the class-based division of work emphasises the diversity of women's experiences of new technologies and specific modes of delivery. Furthermore, the analysis highlighted that perceptions of modern hospital deliveries and interventionist practices are not necessarily the result of middle-class consumption patterns and the emergence of private healthcare, although this and health insurance play an important role (Lazarus 1997, Stivens 1998, Taylor 2000, Jacobson 2001).

The material suggests that even relatively privileged young women experience childbirth as crucially embedded in the way kin relations and households are organised, and that their knowledge of intra-household hierarchies as well as the division of labour in the house and between women belonging to different backgrounds, determines their approach to medicalised childbirth and specific types of deliveries. Pregnancy and birth underwent multiple changes and here, as elsewhere, the medicalisation of childbirth led to the marginalisation of forms of knowledge located in the domestic domain (Ginsburg and Rapp 1991: 321–23). But in the current context this process took a different form than one might expect. It is apparent from the interviews and the two case studies that young birthing women have not become passive recipients but negotiate their relative well being within a context, where hospital births and interventionist deliveries seem to enhance their control over the birthing process. If their prioritisation of rest, support from relatives and their ideal of a painfree delivery outside the home are interpreted in the context of their domestic roles as daughters-in-law, medical knowledge, hospital births and indeed Caesarean sections seem to fulfil multiple needs.

Women themselves learnt how to gain from the increasing medicalisation of reproduction, albeit at the price of subjecting their bodies to various interventionist procedures. Maternal histories were earlier dominated by the shift from birthing as something experienced by daughters in their parents' house towards the gradual subordination of the birthing woman under the strict regime of her parents-in-law. Backed up by their moderate, but increasing economic independence from their parents-in-law, they construct birthing through reference to knowledge generated within the domestic sphere and the authority of medical practitioners situated outside the domestic sphere to negotiate the circumstances.[21] The role of a patient helps to partly transcend the limitations of the role of a daughter-in-law, and although the standard of care provided is often far from satisfactory, a growing number of women enjoy improved facilities available in the private sector. Thus, where critical perspectives on hospital deliveries and 'medicalised' childbirth emphasise that women must experience an extreme loss of control and alienation in the course of this procedure, these women insist that they actually gained some degree of control and choice.

Middle-class Bengali women identify strongly with the kind of healthcare provided during pregnancy and birth, and the distinctive value of hightech Caesarean deliveries adds to their popularity (Bourdieu 1984). I would nevertheless argue that, while different modes of childbearing – and by extension approaches to reproductive technologies – are clearly related to class-based identities, they are not merely a reflection of economic standing or access to specialist services in urban areas. In the given setting the workings of gender and class determine in a more subtle way, whether a woman can afford to forego the allegedly faster restoration of physical strength attributed to 'normal' birth. Where birthing women can today be treated as patients who undergo an 'operation', they are entitled to certain privileges, like extended periods of rest and state-of-the-art medical treatment. Supported by the legitimising discourse of medicalised childbirth, they can voice preferences, expect assistance and negotiate specific aspects of antenatal and postpartum care. This explains why young women experience the new mode of delivery as empowering, and a limited – but arguably growing – sense of agency.

Notes

Fieldwork was conducted between September 1995 and April 1997, and between October 1999 and August 2000 and supported by the ESRC and the Research Fund of the University of London. The argument put forward here is further explored in an article in *Medical Anthropology* 22: 4 and I am particularly grateful to their anonymous reviewers and the editor for many useful comments. The material has been presented in the course of seminars at the London School of Economics, the University of Sussex, and the International Convention of Asian Scholars in Berlin. The suggestions made by participants on all occasions were very helpful and I am also indebted to Chris Fuller, Deborah James, Geert De Neve and Hendrik Wittkopf, who went through drafts of this chapter, as well as Roseanna Pollen and Ursula Frenzel for information on medical aspects.

1. The second National Family Health Survey carried out between 1998–1999 states that one-third of all births take place in health institutions, especially in urban areas where 34 percent of all births take place in such environments. Out of the overall rate of births in health institutions, half occur in private hospitals and the like (International Institute for Population Sciences (IIPS) and ORC Macro 2002: 294–96). The same report states that it is estimated that during the three years preceding the survey 7 percent of all deliveries in India were Caesarean sections (International Institute for Population Sciences (IIPS) and ORC Macro 2002: 299) 'Elective' Caesarean sections are popular in many countries, notably the U.S. and across Latin America, where a correlation between higher national products, a well developed private health sector, and rising rates of Caesarean sections has been established (Belizán et al. 1999, Potter et al. 2001).
2. The same holds true for the growing body of comparative literature dedicated to the cross-cultural study of the ways Western models of medicalised childbirth and new technologies are producing new forms of 'authoritative knowledge' (Sargent 1989; Jordan (1978) 1993: 152ff; Davis-Floyd and Sargent 1997; Ram and Jolly 1998).
3. In colonial discourse the 'domestic' became synonymous with the 'native' sphere, and the emerging nationalist movement constructed the domestic, local and feminine as a powerful source of female power (*shakti*) (see Chatterjee 1993).
4. MaMuaris originating from Western India.
5. Rates of female employment in West Bengal are notoriously low. According to the 1991 Census, Calcutta had the lowest rate of women workingin the organised sector, where out of the overall 30.36 percent employees only 5.42 percent were female. Rates of female employment in this sector in other Metropolitan Areas are: Delhi 7.46 percent, Madras 8.46 percent and Greater Bombay 10.32 percent (Census of India 1991: 13).
6. She is referring to legislation issued to prevent sex-discrimination tests and the selective abortion of female foetuses.
7. Like many middle-class mothers she did not attempt to breastfeed.
8. The strategies employed often included abortion (Jolly 1998).
9. Like in the Madras Presidency, Anglo-Indians and Bengali Christians in Calcutta were the first to accept the services on offer (Van Hollen 1998).

10. Usually doctors, not the management or nursing staff, are held responsible if procedures are unsuccessful.
11. According to doctors with work experience here and abroad, routine antenatal care in these institutions matches that provided, for instance, in Germany.
12. These are ceremonies during which the mother-to-be is served 'desired food' and worshipped as auspicious and fertile by the women belonging to her husband's and/or her own patriline. Rituals marking the 'ripening/hardening' (*pakka*) of the child's body in the womb are found all over northern India and it is common to invite friends and family.
13. As in-married wives, they had few opportunities to get to know the 'traditional birth attendant'. Unlike the Rajasthani villagers studied by Patel, none of the women mentioned special attendants for different castes or a non-commercial relationship with *dais* (Patel 1998).
14. Until as late as the 1960s doctors called to pregnant patients diagnosed while the woman in question was represented by a male member of the family or hidden behind a screen.
15. It should be noted that the occupation of a nurse carries negative connotations in this setting.
16. An important aspect is the arbitrariness of the decisions taken by the *dais*, which have also been studied by Rozario in rural Bangladesh (Rozario 1998).
17. The multiple meanings of pain cannot be discussed here, but it is remarkable that women never refer to any strengthening propensity of pain endured during birth.
18. Just how powerful the image of vaginal deliveries as painful is, can be discerned from fact that the expression *baeta* used to refer to contractions is the generic term for pain.
19. According to the Human Development Report published by the UN the infant mortality rate in India was 71 per 1000 live births in 1997 and maternal mortality was stated as 570 per 100 000 live births in 1990 (United Nations Human Development Report 1999). These rates compare negatively with countries like Thailand, where the infant mortality rate was 31 per 1000 live births in 1997 and maternal mortality was 200 per 100 000 births in 1990. Though the maternal mortality rates decreased marginally in the meantime, maternal deaths in India account for more than 20 percent of the worldwide maternal mortality rate.
20. The term labour power is used in the sense of women's contribution to domestic labour, which as Sangari asserts is 'culturally specific and changing' and 'is structured by social institutions, the labour market, and by other institutions that regulate labour and inheritance'. In the given setting arranged marriages, patrilocality and middle-class domesticity provide the basis for the exploitation of a daughter-in-law's labour power (Sangari 1999: 279).
21. GPs working with Bengali women in London note that they often prefer a non-Bengali doctor, as they are allegedly less likely to treat a patient as a 'daughter' or 'daughter-in-law' (Roseanna Pollen, personal communication).

References

Belizán, J. M., Althabe, F., Barros, F. C. and Alexander, S. 1999. 'Rates and Implications of Caesarean Sections in Latin America: Ecological Study'. *British Medical Journal*, 319: 7222, 1397–400

Béteille, A. 1996. 'Caste in Contemporary India', in C. J. Fuller, ed. *Caste Today*. Delhi: Oxford University Press, 150–79

Borthwick, M. 1984. *The Reluctant Debutante: The Changing Role of Women in Bengal 1875–1927*. Princeton: Princeton University Press

Bose, N. K. 1968. *Calcutta 1964: A Social Survey*. Bombay: Lalvani Publishers

Bourdieu, P. 1984. *Distinction: A Social Critique of the Judgement of Taste*. London: Routledge

Census of India. 1991. Published by the Office of the Registrar General, India

Chatterjee, P. (1992)1995. 'A Religion of Urban Domesticity: Sri Ramakrishna and the Calcutta Middle Class', in P. Chatterjee and G. Pandey, eds. *Subaltern Studies VII: Writings in South Asian History and Society*. Delhi: Oxford University Press, 40–68

―――― 1993. *The Nation and Its Fragments: Colonial and Postcolonial Histories*. Princeton: Princeton University Press

Davis-Floyd, R. E. and Sargent, C. F., eds. 1997. *Childbirth and Authoritative Knowledge: Cross-Cultural Perspectives*. Berkeley: University of California Press

Dickey, S. 2000. 'Permeable Homes: Domestic Service, Household Space, and the Vulnerability of Class Boundaries in Urban India'. *American Ethnologist*, 27: 2, 462–86

Donner, F. H. 1999. *Women and Gold: Gender and Urbanisation in Contemporary Bengal*. Ph.D. thesis, Department of Anthropology, London School of Economics and Political Science

Dwyer, R. 2000. *All you want is Money, all you need is Love: Sexuality and Romance in Modern India*. London: Continuum

Edwards, J., Franklin, S., Hirsch, E., Price, F. and Strathern, M. 1993. *Technologies of Reproduction: Kinship in the Age of Assisted Conception*. London: Routledge

Engels, D. 1996. *Beyond Purdah? Women in Bengal 1890–1939*. Delhi: Oxford University Press

Fernandes, L. 2002. 'Rethinking Globalization: Gender and the Nation in India', in M. de Koven, ed. *Feminist Locations: Global/Local/Theory/Practice in the Twenty-First Century*. New Brunswick: Rutgers University Press, 147–67

Georges, E. 1997. 'Fetal Ultrasound Imaging and the Production of Authoritative Knowledge in Greece', in R. E. Davis-Floyd and C. F. Sargent, eds. *Childbirth and Authoritative Knowledge: Cross-Cultural Perspectives*. Berkeley: University of California Press, 91–112

Ginsburg, F. D. and Rapp, R. 1991. 'The Politics of Reproduction'. *Annual Review of Anthropology*, 20: 3, 11–43

―――― 1995. 'Introduction: Conceiving the New World Order', in F. D. Ginsburg and R. Rapp, eds. *Conceiving the New World Order: The Global Politics of Reproduction*. Berkeley: University of California Press, 1–17

Handwerker, L. 2002. 'The Politics of Making Modern Babies in China: Reproductive Technologies and the "New Eugenics"', in M. C. Inhorn and F. van Balen, eds. *Infertility around the Globe: New Thinking on Childlessness, Gender, and Reproductive Technologies*. Berkeley: University of California Press, 298–314

Hunt, N. R. 1999. *A Colonial Lexicon of Birth Ritual, Medicalisation, and Mobility in the Congo*. Durham and London: Duke University Press

Inhorn, M. C. 2002. 'Introduction. Interpreting Infertility: A View from the Social Sciences', in M. C. Inhorn and F. van Balen, eds. *Infertility around the Globe: New Thinking on Childlessness, Gender, and Reproductive Technologies*. Berkeley: University of California Press, 3–32

International Institute for Population Sciences (IIPS) and ORC Macro. 2002. *National Family Health Survey (NFHS-2): India 1998–99*. Delhi: Mumbai: IIPS

Jacobson, N. 2001. 'Critical Perspectives: The Feminist Critique of Medicine, Medicalization, and the Making of Breast Implant Policy', in C. M. Obermeyer, ed. *Cultural Perspectives on Reproductive Health*. Oxford: Oxford University Press, 221–29

Jeffery, P., Jeffery, R. and Lyon, A. 1989. *Labour Pains and Labour Power: Women and Childbearing in India*. London: ZED Books

Jolly, M. 1998. 'Introduction: Colonial and Postcolonial Plots in Histories of Maternities and Modernities', in K. Ram and M. Jolly, eds. *Maternities and Modernities: Colonial and Postcolonial Experiences in Asia and the Pacific*. Cambridge: Cambridge University Press, 1–25

Jordan, B. (1978) 1993. *Birth in Four Cultures: A Crosscultural Investigation of Childbirth in Yucatan, Holland, Sweden, and the United States*. Prospect Heights: Waveland Press

Lazarus, E. 1997. 'What Do Women Want? Issues of Choice, Control, and Class in American Pregnancy and Childbirth', in R. E. Davis-Floyd and C. F. Sargent, eds. *Childbirth and Authoritative Knowledge: Cross-Cultural Perspectives*. Berkeley: University of California Press, 132–58

Martin, E. 1987. *The Woman in the Body: A Cultural Analysis of Reproduction*. Boston: Beacon Press

Mukherjee, S. N. 1993. *Calcutta: Essays in Urban History*. Calcutta: Subarnarekha

Nair, P. T. 1990. 'The Growth and Development of Old Calcutta', in S. Chaudhuri, ed. *Calcutta: The Living City*. Vol. I, Delhi: Oxford University Press, 10–23

Pai, M., Sundaram, P., Radhakrishnan, K. K., Thomas, K. and Muliyil, M. P. 1999. 'A High Rate of Cesarian Sections in an Affluent Section of Chennai: Is it a Cause for Concern?' *National Medical Journal of India*, 12: 4, 156–58

Papanek, H. (1979)1989. 'Family Status-Production Work: Women's Contribution to Social Mobility and Class Differentiation', in M. Krishnaraj and K. Chachana, eds. *Gender and the Household Domain: Social and Cultural Dimensions*. Delhi: Sage, 97–115

Patel, T. 1998. *Popular Culture and Childbirth: Perceptions and Practices in Rural Rajasthan*. Paper presented at the Centre for Cooperative Research in Social Science, Pune

Patel, V. 1989. 'Sex-Determination and Sex-Preselection Tests in India: Modern Techniques for Femicide'. *Bulletin of the Concerned Asian Scholar*, 21: 1, 2–10

Potter, J. E., Berquo, E., Perpetuo, I. H. O., Leal, O. F., Hopkins, K., Souza, M. R. and de Carvalho Formiga, M. C. 2001. 'Unwanted Caesarean Sections among Public and Private Patients in Brazil: Prospective Study'. *British Medical Journal* 323: 7322, 1155–58.

Raheja, G. G. and Gold, A. G. 1994. *Listen to the Heron's Words: Reimagining Gender and Kinship in North India*. Berkeley: University of California Press

Ram, K. and Jolly, M., eds. 1998. *Maternities and Modernities: Colonial and Postcolonial Experiences in Asia and the Pacific*. Cambridge: Cambridge University Press

Riessman, C. K. 2002. 'Positioning Gender Identity in Narratives of Infertility: South Indian Women's Lives in Context', in M. C. Inhorn and F. van Balen, eds. *Infertility around the Globe: New Thinking on Childlessness, Gender, and Reproductive Technologies.* Berkeley: University of California Press, 152–70

Rozario, S. 1998. 'The *Dai* and the Doctor: Discourses on Women's Reproductive Health in Rural Bangladesh', in K. Ram and M. Jolly, eds. *Maternities and Modernities: Colonial and Postcolonial Experiences in Asia and the Pacific.* Cambridge: Cambridge University Press, 144–76

Sangari, K. 1999. *The Politics of the Possible: Essays on Gender, History, Narrative, Colonial English.* Delhi: Tulika Press

Sargent, C. F. 1989. *Maternity, Medicine and Power: Reproductive Decisions in Urban Benin.* Berkeley: University of California Press

Sesia, P. M. 1997. '"Women come here on their own when they need to": Prenatal Care, Authoritative Knowledge, and Maternal Health in Oaxaca', in R. E. Davis-Floyd and C. F. Sargent, eds. *Childbirth and Authoritative Knowledge: Cross-Cultural Perspectives.* Berkeley: University of California Press, 397–420

Standing, H. 1991. *Dependence and Autonomy: Women's Employment and the Family in Calcutta.* London: Routledge

Stephens, M. 1986. 'The Childbirth Industry: A Woman's View', in L. Dube, E. Leacock and S. Ardener, eds. *Visibility and Power: Essays on Women in Society and Development.* Delhi: Oxford University Press, 70–84

Stivens, M. 1998. 'Modernizing the Malay Mother', in K. Ram and M. Jolly, eds., *Maternities and Modernities: Colonial and Postcolonial Experiences in Asia and the Pacific.* Cambridge: Cambridge University Press, 50–80

Szurek, J. 1997. 'Resistance to Technology-Enhanced Childbirth in Tuscany: The Political Economy of Italian Birth', in R. E. Davis-Floyd. and C. F. Sargent, eds. *Childbirth and Authoritative Knowledge: Cross-Cultural Perspectives.* Berkeley: University of California Press, 287–314

Tatar, M., Günalp, S., Somunoglu, S. and Demirol, A. 2000. 'Women's Perceptions of Caesarean Sections: Reflections from a Turkish teaching hospital'. *Social Science and Medicine,* 50: 9, 1227–33

Taylor, J. 2000. 'Of Sonograms and Baby Prams: Prenatal Diagnosis, Pregnancy, and Consumption'. *Feminist Studies,* 26: 2, 391–418

United Nations 1999. *Human Development Report 1999*

Unnithan-Kumar, M. 2002. 'Midwives among Others', in S. Rozario and G. Samuel, eds. *Daughters of Hariti. Childbirth and Female Healers in South and Southeast Asia.* London: Routledge, 27–52

Van Hollen, C. C. 1998. 'Birthing on the Threshold: Childbirth and Modernity among Lower Class Women in Tamil Nadu, South India'. Ph.D. thesis: University of California, Berkeley

Varma, P. K. 1998. *The Great Indian Middle Class.* Delhi: Penguin Books

Walsh, J. 1995. 'The Virtuous Wife and the Well-Ordered Home: The Reconceptualisation of Bengali Women and their Worlds', in R. K. Ray, ed. *Mind, Body and Society: Life and Mentality in Colonial Bengal.* Calcutta: Oxford University Press, 331–59

CHAPTER 7

IN SEARCH OF CLOSURE FOR QUINACRINE: SCIENCE AND POLITICS IN CONTEXTS OF UNCERTAINTY AND INEQUALITY

Asha George

Introduction

Over 100,000 women have been chemically sterilised using Quinacrine during the past thirty years in twenty countries (Kessel 1996), yet the scientific and ethical basis of this practice remains contentious. A diverse group of actors involving activists, doctors and philanthropists, as well as advocacy, research and funding organisations, have mobilised around this controversy. My research[1] analyses how these groups both construct authoritative knowledge and mobilise support to achieve their regulatory goals.

Quinacrine's current controversial status is part of a history of tension between scientists, policy makers and activists with respect to contraception. Divergent ideologies concerning population control, reproductive health and rights distinguish these groups. They conflict in their assessments about the safety, efficacy and effects of contraceptive methods, as well as disagree about the influence different societal contexts have on the development and distribution of contraceptive methods (Hardon 1992, Ravindran and Berer 1994). Although these disagreements vary depending on the particular contraceptive in question, the balance of power between scientists and activists tends to tilt towards the former rather than the latter. Nonetheless, Quinacrine shows how marginalised groups can still achieve substantial influence against status quo positions.

I explore how Quinacrine critics, initially perceived as radical and non-expert, succeeded in reversing policy to their benefit by recasting

Quinacrine promoters as non-scientific extremists. My analysis is based on qualitative field research and a review of social theories about expertise and mobilisation. Although conventional perspectives tend to dichotomise the policy success of activists to mobilisation tactics and that of scientists to technical expertise, in actuality both activists and scientists use similar tactics to influence policy to their benefit. This leads us to reconsider policy models conventionally used for resolving controversies in the field of contraceptives, as well as in public health more broadly. Rather than separate science from politics, favouring one above the other, I conclude that regulatory efforts must emphasise a more democratic basis for research, which skilfully combines rather than separates science, politics, ethics and objectivity.

Case Study: Quinacrine Sterilisations

Initial Trials

Quinacrine was originally used in the 1920s as an anti-malarial drug. Dr Jaime Zipper started experimenting with it as a sterilising agent on Chilean women during the 1960s and 1970s (Zipper et al. 1970, Zipper et al. 1975). He found that by using a modified intra-uterine device (IUD) inserter, he could deposit Quinacrine pellets in a woman's uterus.[2] The pellets melt into Quinacrine hydrochloride, a burning acid causing scar tissue. This scar tissue gradually blocks the fallopian tubes leading to sterilisation.

In contrast to Chile, research in the U.S. started in the late 1970s with animals rather than with women. Apart from dose-related toxic effects, researchers also found a positive Ames test result indicating that Quinacrine is mutagenic in bacterial models (Pollack and Carignan 1993). Despite these worrying results, a literature review showed no clear evidence of changes in mammalian DNA. Those who defend Quinacrine argue that the information available from these first studies provide no cause for concern about the longterm safety of Quinacrine in humans (Kessel 1997).

Subsequently a World Health Organisation (WHO) Toxicological Panel reviewed the use of Quinacrine installations. The panel concluded that further animal studies were required to evaluate its safety before clinical trials on women could begin (Benangiano 1994). Doubts about the safety of the method strengthened when an abnormal association of cancer among the Chilean women prompted Family Health International (FHI) to discontinue support for Zipper's work in 1989 (Interviewee No. 9). Although analysis of retrospective data suggested that there was no increased risk of cancer, FHI cautioned that the study sample was too small to prove Quinacrine's safety (Network 1995).

Despite these conclusions by WHO and FHI, trials continued to take place on women in low-income countries. This fact became glaringly apparent when *The Lancet* published the results of a trial conducted by the

Vietnamese government involving 31,781 women (Hieu et al. 1993). Criticism focused on methodological issues of study design and data analysis. Concern also focussed on the possibility that sterilisations were done without adequate informed consent or even knowledge of the women involved (AVSC 1993, Mulay 1997).

The Association for Voluntary Surgical Contraception[3] (AVSC) distributed a technical statement assessing the scientific basis for Quinacrine sterilisations and hosted an expert meeting attended by various organisations, including women's groups. Without claiming to represent a consensus the meeting report reaffirmed the 1991 WHO position. No further clinical trials or regular use in women should take place until toxicological studies in animals and retrospective studies were completed (Carignan et al. 1994). Subsequent pressure from WHO was applied to stop Quinacrine trials by government researchers (*The Lancet* 1994, Mumford 1994).

Despite excitement about the possible discovery of a better method of sterilisation, doubts about longterm cancer risks and efficacy of the method could not be quelled. Information about higher rates of ectopic pregnancies, side effects, and the potential for abuse caused many public health organisations and women's groups to question the continued support for Quinacrine as a sterilisation method. Since current scientific understanding on Quinacrine rests on inconclusive evidence, faulty methodological foundations and troublesome ethical experience (AVSC 1993; Mulay 1997), some critics argue that Quinacrine should be discontinued from the contraceptive research agenda (Interviewee No. 13).

Indian Trials

Nonetheless, clinical trials on poor women continued to take place in India. Based on their own prior research private doctors were able to prompt the Indian Council for Medical Research (ICMR) to conduct a limited trial in 1992. This trial was discontinued after studying only eight women because of a 50 percent failure rate and the development of fever in two cases (The Hindu 1997). Since then the Drug Controller of India (DCI) and ICMR repeatedly state that they did not authorise the use of Quinacrine as a sterilising agent.

Yet a national network of primarily private doctors,[4] involving prominent gynaecologists revered by the medical community, continued to do exactly that. The most influential among them is Dr Jain. A previous Bharatiya Janata Party (BJP) Member of Parliament, Jain leads various private medical ventures. These include medical and public health services, professional associations, broadcasting and publishing enterprises. Jain also serves as the President of the International Federation of Family Health (IFFH) and thus has a formal link to Dr Elton Kessel.

Dr Elton Kessel (former director of FHI)[5] and Dr Stephen Mumford (also formerly employed by FHI) supported Zipper's research on Quinacrine sterilisations. Since their departure from FHI, Kessel and Mumford continue to promote the use of Quinacrine worldwide through medical networks and their own respective non-governmental organisations (IFFH and the Centre for Research on Population and Security).

Other key Quinacrine promoters in India include Dr Pravin Kini and Dr Sita Bhateja.[6] They created the Contraceptive and Health Innovations Project (CHIP Trust) in Karnataka. In addition, Dr Biral Mullick, President of the Indian Rural Medical Association, started doing Quinacrine sterilisations in the 1970s. He states that over 10,000 Quinacrine procedures have been carried out in West Bengal since then. He also claims to have instructed 300 rural doctors in Quinacrine insertion through their basic training in Community Medical Service with the Humanity Association (*Quinacrine Sterilisation Newsletter* 1997: 2).

These illegal activities did not go unnoticed by either the women's movement or the public health community in India. On 16th March 1998, the Supreme Court of India heard the Public Interest Litigation case filed by the All India Democratic Women's Association (AIDWA) and the Centre for Social Medicine and Community Health at the Jawaharlal Nehru University (JNU) against the DCI, Jain Clinics and the CHIP Trust. In its affidavit the DCI notified the Supreme Court that it was in the process of issuing a ban against Quinacrine sterilisations. Another six months were to pass before the DCI issued the ban on 17th August 1998.

Constructing Scientific Expertise and Authoritative Knowledge

Although Quinacrine is currently regarded by some as an extreme and bizarre case in contraceptive research, this research initially continued for a period of roughly thirty years (1960s–1990s) without sparking public controversy. Once it became controversial, research continued irrespective of government regulation, a technical policy articulated by WHO and activist outrage. This persistence of Quinacrine promoters led me to examine why certain kinds of expertise are accepted, by whom and how such acceptance allows certain beliefs to be normalised beyond introspection. How did Quinacrine promoters and critics manage to mobilise support and persist in their actions against the face of adversity?

The literature on epistemic communities and transnational advocacy networks points to causal beliefs as forming a crucial element in justifying claims, as well as in mobilising supportive audiences (Haas 1992, Keck and Sikkink 1998). In addition, social constructivists have also analysed how scientists and other groups seek to consolidate and guard their authority while subverting that of others (Gieryn 1999). I follow Gieryn's analysis of scientific controversies to describe the policy processes through which boundaries are negotiated to secure authoritative knowledge about Quinacrine.

Causal Beliefs Motivating Claims and Audiences

The overriding concern of Quinacrine promoters is the imperative of population control. 'The population explosion is India's number one problem'

(Jain undated). Moreover 'population control is an undisputed national objective for obvious, social, economic and environmental reasons' (Kini undated). Apart from being a national priority in India, population pressure also has international consequences as it drives illegal immigrants from developing countries to the United States (Interviewee No. 17). For this reason a growing population is described as 'runaway', 'overwhelming' and causing 'unsustainable' 'demographic damage'. Family planning workers are called 'heroic' and doctors are 'highly qualified, patriotic and conscientious' for being concerned with the problem of population (Kini undated, Kini et al. undated, Jain correspondence).

Quinacrine promoters believe that female sterilisations empower women and address maternal mortality. 'Every 100 sterilisation (*sic*) prevent one young mother from dying, and three young children from being orphaned' (Kini undated). They argue that as sterilisation is the most used method, it is 'the most popular method of contraception in India and world-over' and 'demanded' by women (Kini undated). This massive demand for female sterilisation causes health providers to compromise their quality of care and increase the risks of surgical sterilisation. Quinacrine, in contrast to surgical sterilisation, is cheaper and easier to use. It is therefore especially 'relevant for developing countries like India, where surgeons and surgical facilities are not available to deliver surgical sterilisation services to our mothers, specially those who live in rural and tribal areas' (Jain undated).

Quinacrine is therefore seen to be *the* technical solution to the social tragedies of poverty and parity borne by poor, rural and slum women in India. This reductionist view of social problems and faith in scientific and technological progress discounts efforts to improve medical services, encourage the use of different kinds of contraceptive options, address poverty and promote gender equality.

Furthermore, Quinacrine promoters perceive a comparative advantage in undertaking contraceptive research on women in low-income countries. The benefits of such research will be higher in low-income countries, because of higher maternal mortality rates, rapid population growth and high poverty levels (Kessel 1996).[7] As a result, they argue that the standards for research should also be more flexible in low-income countries. 'When it comes to contraceptive research, the North has the resources and the technical capability to conduct preclinical toxicological trials that the South can ill afford, while the South can more economically document efficacy, acceptability and early safety in trials of a new contraceptive method' (Kessel 1994). Quinacrine promoters believe that the theoretical possibility of longterm cancer risks cannot be determined without research by Indian doctors of widespread use by poor women (Jain undated). Indeed Indian doctors must get involved as pharmaceutical companies will not facilitate drug research or distribution as Quinacrine cannot be patented.[8]

In conclusion, Quinacrine promoters present themselves as altruistic public crusaders. These brave doctors are single handedly spreading the benefits of modern drugs to poor Indian women, protecting these mothers from repeated pregnancies and maternal mortality, as well as saving India and the world from the ravages of increasing population and poverty. These beliefs allow them to persist flouting ethical standards and shield them from scientific and political adversity. The legitimacy of their research enterprise is further bolstered by their construction of authoritative knowledge as detailed in the next section.

Consolidating Authoritative Knowledge

Among medical doctors, awareness of Quinacrine is spread through networking at technical forums and seminars that help to give an impression of scientific legitimacy. Kessel and Mumford, as the central nodes of the network, provide technical and material support to interested doctors in low-income countries. Through the Institute for Development Training, they produce the *Quinacrine Sterilisation Newsletter*, the QS website, instructional videos, information pamphlets and compilations of their publications. Similarly the CHIP Trust in Karnataka, mailed literature on Quinacrine to members of the Obstetric and Gynecology (OB/GYN) society and organised specific sessions on Quinacrine at medical congresses (Kini 1996). Facts[9] and study results about Quinacrine are published in 'scientific' journals[10] whose editors consist of fellow Quinacrine promoters.

Since Quinacrine was used for the treatment of malaria, giardia and some other clinical conditions, Quinacrine promoters consider it to be a safe and acceptable drug for sterilisation (Kessel 1994). Although negative effects of its sterilisation use are reported, doctors consider these to be secondary, temporary and minor. Risks are acknowledged, but doctors are more qualified to rationally determine their assessment.[11] Indeed the disciplinary training and scientific expertise of these doctors is interpreted as an endowment of authority independent of government regulation. Jain quotes a 1982 U.S. Federal Drug Agency bulletin authorising doctors to use drugs for off-label purposes as, 'experience demonstrates that the official label lags behind scientific knowledge and publications [...] Thus, the official government approvals for this use of Quinacrine may be sought if one wants to, but not necessary' (Jain 1997).

Contrary to governmental denial of knowledge about Quinacrine use in India, several doctors had presented their Quinacrine research to the government in order to advise the Toxicology Review Meeting held on 17 June 1992 in preparation for the ICMR trials. Jain was also subsequently involved in various toxicology and ethical review meetings at ICMR and at the Ministry for Health and Family Welfare (Interviewee No. 12). The CHIP Trust also had access to government officials within the State Government of Karnataka, and was at one point waiting for authorisation from the central government to proceed with Quinacrine use in government clinics in 1994 (Mendonca 1994, Mumford 1994). Lastly, the delay in acting on the Drug Technical Advisory Board's decision to ban Quinacrine sterilisations in India from December 1997 until August 1998 may be because Jain personally lobbied the DCI and the Ministry of

Health and Family Welfare to order a review of the DTAB decision (Interviewee No. 12).

To conclude, Quinacrine promoters advance their expertise through professional medical networks that were able to infiltrate government bodies, including those committees that are supposed to ensure the ethical review of research protocols. In India these biomedical networks and committees are not easily accessible to the public. Making technical or ethical review meetings publicly accessible is not only a challenge for India. The Planned Parenthood Federation of America's (PPFA) board meetings gathered information primarily from Quinacrine promoters, with less attention paid to information from biomedical experts who were more critical of Quinacrine sterilisations or from respected women's health advocates (Interviewee No. 23). Not only do these professional exchanges require an understanding of medical terminology, but they also take place in a biomedical social milieu that is rarely, if ever, populated by feminist activists, critical public health advocates, social science analysts or even by women users themselves. Technical expertise and consensus is achieved in this case by assuming public authority, while excluding contradictory evidence and critical voices.

Subverting Competing Knowledge

When there is disagreement within the scientific community about the interpretation of evidence, Quinacrine promoters quickly state that certain kinds of facts and expertise matter more than others. With respect to the ICMR trial, they choose to highlight that six out of the eight sterilisations succeeded. They feel that the government erroneously focused on the two sterilisations that failed (Interviewee No. 12).[12] When the DTAB recommended the banning of Quinacrine in India, Quinacrine promoters responded by lobbying for the creation of a sub-committee of more appropriate experts. This sub-committee would need the representation of obstetric gynaecologists, human reproduction scientists, or any Indian researcher with an 'understanding of the needs of overwhelmingly rural, often poor women' (Jain undated). Pharmacologists and general practitioners were considered by Quinacrine promoters to be not appropriately qualified to decide such matters for the DTAB.

Quinacrine promoters draw boundaries to distinguish between correct and incorrect scientific practice in order to suit their goals. Several Quinacrine supporters separate good science from value judgements and chastised WHO for its espousal of universal research ethics. 'In applying a North standard for contraceptive research to a developing country, HRP has made a value judgement that is not based on science or logic' (Kessel 1994).

Lastly, as highly qualified and innovative medical researchers, Indian Quinacrine promoters are already frustrated by the lack of opportunities for advancing their careers in India. 'India is a free country and any kind of mental slavery to the notion that wisdom flows only from western minds cannot be acceptable to Indian researchers, who are trying to do their best to find solutions to medical problems which confront Indian society in our socio-economic conditions' (Jain undated). Indians who are critical of the method are 'elite groups influenced by western culture who might not have the understanding or the concern for the socio-economic conditions in which a majority of the women of our country live' (Jain undated). These themes find a welcome reception[13] in the rise of Hindu fundamentalism marked by the political ascendancy of the nationalist BJP in Indian politics.

Guarding Authoritative Knowledge

Yet when presented with incriminating issues that cannot be evaded, Quinacrine promoters guard their credibility by withdrawing the boundaries of their expertise and influence. They agree to the need for following up patients, but excuse themselves for lack of resources, lack of jurisdiction over another doctor's professional autonomy or lack of feasibility considering the social context from which many women using Quinacrine came (poor, migrant or urban slum women). When confronted with the reality of the weakness of peer review and self-regulation within the Indian medical community, Quinacrine promoters regard this and the admitted breaches in research protocols with dismay. These problems are accepted as being natural when 'doctors can get away with anything in India' and thus not worthwhile pursuing (Interviewee No. 8, 12).

The private medical doctors involved can afford to be nonchalant and aloof about their lack of engagement with these problematic issues, because of their position of power and privilege in Indian society. Unlike government researchers, they are not held accountable to effective peer review mechanisms, nor are their employers vulnerable to public inquiry from parliament or international pressure from inter-governmental organisations like WHO. Most importantly, they are many times more socially and economically powerful than the poor, illiterate women they have sterilised with this experimental procedure. Nonetheless the case of Quinacrine shows that even powerful, well established groups can be destabilised.

Creating Feminist Alternatives

Causal Beliefs Consolidating Change

Quinacrine critics succeeded in legitimising their concerns about Quinacrine sterilisations, because their policy strategies included constructing alternative forms of expertise hinged on certain causal beliefs. Many critics see Quinacrine as an example of the demographic focus of contraceptive research. This leads to a concern for efficacy to the neglect of other important issues like safety, acceptability and the potential for abuse of a method. Feminists argue that scientists neglect side effects, unknown long term cancer risks, high prevalence of comorbidities like reproductive tract infections and anaemia in women, and the effects on women's sexual desire, when exploring new contraceptive methods (Weiringa and Hardon 1993).

These feminist concerns were able to consolidate a landmark consensus at the 1994 International Conference on Population and Development in Cairo after which family planning and population issues could no longer be solely seen from a demographic perspective. Not only did the individual rights of women as users of such services become important, but the primary goals of such services were to promote the comprehensive health and well being of their clients. With this change it became professionally indefensible to promote contraceptive methods that reduced fertility at the risk of women's health and that would further circumvent the provision of health services to poor women.

Social context and the quality of health services are deemed as essential requisites for appropriate contraceptive use and introduction (Spicehandler and Simmons 1994, Simmons et al. 1997). Medical follow up, counselling and emergency care must be available when using complex contraceptives technologies. Indeed Indian women's health activists did organise to ban the use of long acting hormonal contraceptives (Depo-Provera and Norplant) from government programmes, precisely because of the lack of a guarantee of accessible, quality health care (Interviewee No. 3, 9, 13).

Quinacrine critics also argue that the exercise of reproductive rights 'requires broader changes in women's circumstances, not just more and better technology and delivery systems' (Germain undated). Indeed to 'talk of reproductive rights in the face of the lack of right to food, employment, water, access of education, health and indeed even the survival of children – in other words all the accoutrements of survival with dignity – is to make a travesty of women's rights' (Rao 1997). Attention to gender, along with other social relations of power, is critical if one is to enable women's informed use of health services and their ability to hold services and society accountable for their rights (Sen 1995).

Quinacrine's attraction as a low cost, low skill, easily available technology that mimics IUD insertion, is precisely why it can also be used without the consent and knowledge of the women who are sterilised by it. Hence apart from addressing unequal gender relations, vulnerability caused by poverty, broad health and social services, a strong legal and civil society is also necessary to monitor possible cases of abuse (Sen 1995).

Mobilising Support

These alternative beliefs around contraception, reproductive health and rights were promoted through various information outreach activities and consultations. Women's groups created and disseminated a 'Contraceptive Bill of Rights'. HRP/WHO hosted a series of consultations between scientists and women's health advocates called 'Creating Common Ground' to give voice to feminist and user perspectives about contraception.

With respect to Quinacrine, as research initially existed primarily in the biomedical arena, Indian women's groups and public health advocates were unaware of its existence or relevance. The relative silence of the scientific community added to the consensual, legitimate authority of Quinacrine sterilisations. As such, concerned activists and researchers found it initially hard to find receptive audiences (Interviewee No. 9). The coalition of actors that I term as Quinacrine critics[14] actually only came into existence once Quinacrine research became more visible within public health and feminist circles.

Women's groups in India first heard of Quinacrine with the local screening of the BBC documentary 'The Human Laboratory', in which Quinacrine was briefly featured (Interviewee No. 2, 7, 13). Subsequently Saheli distributed an investigative report that reviewed both the scientific debate and the current legislation that was being violated by Quinacrine promoters (Saheli 1997). Women's groups also suggested Quinacrine as a subject of research to media students who created the documentary, 'The Yellow Haze' (Interviewee No. 7).

The subsequent mobilisation of critics through various networks drew support from the high level of activist concern over contraceptive, drug and technology regulation in India. Historic precedence in the form of high dose oestrogen-progestrone (EP) drugs, sex-selective amniocentesis and forced sterilisations has lead to a latent coalition of activists and organisations with significant political experience (Gandhi and Shah 1992). Activists staged protests outside the private clinics of doctors who were using Quinacrine. Questions were raised in parliament and different groups petitioned government officials about the issue. AIDWA and JNU filed a Public Interest Litigation case against the government and the two most prominent Quinacrine promoters in India (Jain and Kini). AIDWA formed a specific group to organise around Quinacrine use in Bangalore. Other women's groups began collaborating with doctors and are subsequently undertaking public health research in the slums of Bangalore (Interviewee No. 6).

However such mobilisation did not lead to immediate policy responses from the government. As Schneider and Ingram (1993) suggest groups that are perceived as negative but powerful can blunt negative policies aimed at them, but it is much more difficult for them to receive visible policy benefits. Policy benefits are usually given discretely as government officials prefer to have public opinion think that they are not favouring such groups. Benefits tend to also to be cloaked in procedures or vague and complex statutes implemented by lower level agencies.

It is not surprising therefore that public officials continued to deny knowledge of Quinacrine use or would term such cases as being extreme aberrations of discredited science that did not merit the level of controversy raised by women's groups. Women's groups were seen to be exaggerating a 'non-issue' (Contact No. 6). Since women's groups in India are negatively perceived, government officials would not risk being seen as responding to them. Due to this inertia activists needed to resort to international links in order to pressure for more moral leverage on the government (Keck and Sikkink 1998).

Statements against Quinacrine by credible international technical organisations (WHO and AVSC) published in international journals (*The Lancet* and *Reproductive Health Matters*) proved to be important resources of credibility during the court case (Interviewee No. 7, 15, 20). Although the Drug Controller of India reported that it was processing the ban order against Quinacrine, six months passed before action finally took place. Activists claim that the publication of a front page *Wall Street Journal* article, a segment on *60 Minutes* among many other national and international press coverage efforts, helped add to the pressure that eventually broke the delay in the public issuance of the ban order in August 1998 (Interviewee No. 3, 13, 20).

It is tempting to believe that the negative press attention brought against Quinacrine promoters by a leading financial and conservative newspaper proved to be so influential. The decision by the Swiss pharmaceutical company to discontinue Quinacrine pellet production, the government ban orders in India and Chile, and lastly the FDA warning letter issued to Kessel and Mumford for their involvement in illegal international drug distribution all took place within weeks of that story. The political costs and benefits of inaction against Quinacrine by public officials had changed. Government could no longer afford to relegate policy to unclear regulative procedures. They had to be seen taking visible action.

Subverting Authoritative Knowledge

Quinacrine critics also deployed public health research that underpinned and followed the Cairo agenda to undermine the knowledge claims made by Quinacrine promoters. For example, the theoretical links between population growth and economic development have shown no consistent universal trend. Moreover the narrow focus on population growth has distracted attention from other issues, such as inequalities in quality of life and consumption patterns between and within countries. Maternal mortality is a prime example of such inequality. Contrary to the claims of Quinacrine promoters, access to contraception alone is not a direct solution to maternal mortality as the pregnancies in question may be wanted or planned pregnancies, in which case these women would not be using contraception in any case (Carignan et al. 1994).

Critics also argue that the emphasis on female sterilisation as a means of ideally expanding contraceptive choice is problematic. First, male sterilisations are easier, safer and cheaper to carry out, hence the rationale for exclusively sterilising women is not clear. Secondly, women's needs are different depending on their current reproductive goals and societal context. In order to provide contraceptive choice one needs access to a range of contraceptive methods throughout one's sexual and reproductive life. Again the exclusive emphasis on sterilisation, or any one method, is inappropriate.

Finally, Quinacrine critics were able to expose fundamental ethical and methodological problems in the research carried out by Quinacrine promoters. This ultimately called into question whether these doctors were actually carrying out research, or using the label of research as a cover for sterilising poor women in low income countries. Not only was the level of informed consent questionable, but doctors did not actively follow up their patients to provide either care or to monitor results. Differing screening, insertion and outcome indicators were used, making it very difficult to compare evidence from different 'trials'. Finally, many of the results were inappropriately generalised from smaller samples to larger patient populations. Apart from abusing ethical issues of accountability of researchers to vulnerable populations, these doctors violated basic principles of clinical research methodology.

By combining the mobilisation of policy support with knowledge construction, Quinacrine critics were able to discredit the seemingly invincible scientific authority of Quinacrine promoters. Not only were Quinacrine promoters ethically shamed, they were also challenged by contrary public health evidence and causal frameworks. Lastly, they were disqualified by the methodological guidelines that govern the field. On these grounds Quinacrine critics were jettisoned from the legitimising realms of science.

In Search of Closure

Many critics assumed that their efforts to mobilise policy and to redraw boundaries around authoritative knowledge were successful, as by now Quinacrine research had come to be associated with fraudulent science. Indeed the series of events in 1998 seemed to indicate decisive policy actions against Quinacrine sterilisations around the world leading to closure. Nonetheless Quinacrine promoters proceeded to plan clinical trials with U.S. women and also organised a workshop promoting Quinacrine sterilisations at the American Public Health Association's Annual Meeting in November 1999. Even more disturbing to women's health advocates was the consideration by the Board of Directors of the Planned Parenthood Federation of America to undertake clinical trials for Quinacrine sterilizations through its various affiliates. In 2001, the Children's Hospital in Buffalo, NY gained FDA approval to undertake human trials (Hartmann 2001). This is being interpreted as giving a green light to those in India, where despite the ban, doctors continue to sterilise women with Quinacrine but charge higher prices (Das Gupta 2001).

Reflecting on these more recent events made me realise that in certain cases consensus based on the exclusion of dissent serves only as a temporary form of closure. Oudshoorn (1994: 125) commenting on the history of the contraceptive pill suggests that, '[m]edical innovation requires the creation of contexts to establish the required links between the technology-in-the-making and its new audiences and consumers. If such a context does not exist, scientists have to create it'. I believe that such linkages cannot be confined to the architecture of scientists alone. Considering the troubling ethical history of contraceptive research and the increasingly unequal worlds that divide scientists and their 'subjects', the creation of stabilising contexts and receptive audiences must welcome the active debate and participation of collectively organised 'subjects'. This approach may be useful in not only redressing abuse, they may also create social and institutional structures that proactively prevent such abuse from taking place in the first place (Sen 1995).

It is useful here to reflect on how Schnieder and Ingram (1993) describe how groups perceived to be positive but weak (poor women) tend to be given benefits that are largely symbolic. Lower level agencies without adequate institutional support or funding are expected to distribute benefits to them that tend to have eligibility requirements, stigmatising labels and are not readily accessible. Governments tend to justify these efforts with welfare or paternalist concerns and the groups are encouraged to be dependent and politically uninvolved. When lower level agencies fail in their tasks, these groups are abandoned to the private sector. Government officials and medical doctors both treat poor Indian women in this manner, and feminists[15] are not immune to similar criticisms.

Although there are instances where poor, rural and slum women in India have been able to collectively represent their own interests, this is still not the case in the arena of contraceptive research. Nonetheless, I argue that in a world of gross inequalities it is unlikely that scientists can by themselves effectively monitor the ethical practice of relevant research without the respectful engagement of these recipients of science and policy. Before exploring this in more detail, I discuss some examples from environmental health to illuminate the nature of two contrasting research and policy models.

Lessons from Environmental Health Regulation

The experience of how science and policy interact in environmental health in the U.S. (Jasanoff 1992) points to interesting parallels for those seeking closure of the Quinacrine controversy. Environmental health regulation, as in the case of contraception, requires safety reassurances in spite of inconclusive or contentious evidence. Apart from protecting healthy people from potential future risks (regulating before harm actually occurred), policy guidelines and the science that underpins them must also resist deconstruction from industry who are required to bear the economic costs of regulation.

Jasanoff describes how in the context of significant scientific uncertainty about longterm cancer risks, the Environmental Protection Agency began to shift away from an emphasis on testable knowledge claims (facts) and began to focus on the process of knowledge production for arriving at factual conclusions (public witnessing). 'The presumption was that the validity of the policy-relevant fact (X's carcinogenicity) could only be established by making explicit the stages leading to its creation'

(Jasanoff 1992: 203). The U.S. Courts supported this focus on *process* rather than *fact* by arguing that regulation could take place on the basis of *substantial but not conclusive evidence* of carcinogenicity. Neither parameters of 95 percent statistical certainty nor medical consensus were required for the Courts to support legislation that sought to protect people's health from longterm risks documented by other types of credible evidence (Jasanoff 1992).

Despite this judicial support, it was only a matter of time for those who disagreed with the EPA's standard setting to question the EPA's process of knowledge production. Clinical and epidemiological studies done by the EPA were faulted for using inadequately validated protocols, omitting proper controls and failing to seek competent peer review. The design of animal studies was questioned as were the principles for extrapolating risk from them. These charges of technical incompetence linked up with accusations of political bias, as the EPA found itself openly agreeing with the scientific positions of environmentalists or regulated industries.

The loss of legitimacy of EPA's science highlights 'the incompatibility between adversarial procedures and the closure of scientific debates relevant to public policy' (Jasanoff 1992: 201). EPA's processes of knowledge production were contested in the same way as laboratory research findings are before they attract substantial alliance from scientists (Latour and Woolgar 1986, Latour 1987). 'Under hostile scrutiny EPA's explanations for its risk determinations became more detailed and explicit, but increasing clarity did not necessarily improve the chances for ending conflict' (Jasanoff 1992: 201). EPA's standards and guidelines, rather than being accepted as starting points, began to be bypassed by alternative scientific models presented by dissenting actors.

Consequently, the second legitimisation strategy pursued by the EPA was to seek council from scientific experts independent from the agency (Jasanoff 1992). Before discussing the reasons for success of this second strategy, I explore the caveats involved in using expert bodies in the following sections. Some controversies cannot be resolved by simply consulting 'neutral' experts and evidence.

Ethics and Objectivity in Scientific Research

Harding discusses two types of politics in science (Harding 1992). The first is more explicit, as groups overtly act to influence the nature of the scientific agenda. Calls for neutrality can adequately safeguard this first type of intrusive politics. The second kind of politics paradoxically works through the depoliticisation of science. Here bias has already infiltrated the institutional structure of science. Rather than providing safeguards for the ethical practice of science, calls for neutrality in this instance further distort the bias inherent in science. The neutrality ideal 'certifies as value-neutral, normal and natural, and therefore not political at all the existing scientific policies and practices through which powerful groups can gain the information and explanations that they need to advance their priorities ... Thus, when sciences are already in the service of the mighty, scientific neutrality ensures "that might makes right"' (Harding 1992: 83).

Harding cautions that by discarding the ideal of neutrality one does not necessarily defile the ideals valued in objectivity, namely fairness, honesty and detachment. She underscores Haskell's argument:

> The very possibility of [...] scholarship as an enterprise distinct from propaganda requires of its practitioners that vital minimum of ascetic self-discipline that enables a person to do such things as abandon wishful thinking, assimilate bad news, discard pleasing interpretations that cannot pass elementary tests of evidence and logic, and, most important of all, suspend or bracket one's own perceptions long enough to enter sympathetically into the alien and possibly repugnant perspectives of rival thinkers. (Haskell 1990: 132 as cited by Harding 1992)

Similarly Jasanoff (1992) notes that under U.S. administrative law regulatory agencies are required to show that they have engaged in *reasoned decision making*. This means that

[a]ssumptions must be spelled out, inconsistencies explained, methodologies disclosed, contradictory evidence rebutted, record references solidly grounded, guesswork eliminated and conclusions supported 'in a manner capable of judicial understanding'. (Rogers 1979 as cited in Jasanoff 1992: 203)

I have chosen these two excerpts to underline my assertion that attention to procedural guidelines are essential for guaranteeing the ethical basis of scientific research. It directly impinges on the quality and credibility of the research undertaken.

Values and bias are inherent in research and policy. One should not try to artificially excise either from research, policy or even objectivity. Rather than letting neutrality monopolise the meaning of objectivity, I argue that we must encourage self-critical consciousness as to where our biases lead us. Most importantly this focus on individual awareness or reflexivity must also be linked to larger collective and institutional mechanisms of accountability. 'Research' is not an activity carried out by individuals alone, whether in private medical clinics or social science research institutes. As the Ethical Guidelines by the Indian National Committee for Ethics in Social Science Research in Health concludes:

> While ethical guidelines are not administrative rules and the conscience of researchers may be the best guide for ensuring that ethics are followed in research and for resolving ethical dilemmas, conduct of research cannot be completely left to the discretion of individual researchers. Institutions and researchers involved in social science research in health should create appropriate institutional or research project based mechanisms to ensure ethical conduct of research and implementation of guidelines. (NCESSRH 2000: 25)

Democracy in Scientific Research

Yet as important as institutional accountability mechanisms are, they are only as good as the people involved in them. Scientists and medical doctors from all over the world may share more in common, via their specialised training and their more privileged social backgrounds, than the research subjects or patients they interact with. For this reason, apart from emphasising individual reflexivity and institutional ethical processes, it is critical to actively seek and engage the perspectives of those less privileged and at the margins of our social worlds.

For example, women's groups call for contraceptive research to be designed and conducted by agencies that are committed to an open research process that must be fully accountable and interdisciplinary in nature. The membership of ethics committees must include women who are critical of the current research paradigm, as well as current and potential users in such numbers as to be able to play a decisive role. Researchers must also be committed to longterm, proactive follow up of all research subjects including detailed morbidity and mortality studies (Canadian Women's Committee on Population and Development 1996, Women's Global Network for Reproductive Rights 1996).

These perspectives coincide with constitutive policy models. Constitutive models emphasise that closure cannot be achieved without reaching shared perceptions and values, which 'arise not from factual [or token] demonstrations but from repeated engagement and consensus building among experts, stakeholders, decision-makers, the media, and the public' (Jasanoff 1998). In order to succeed in this endeavour, emphasis is placed on creating more democratic bases for research that is transparent to all those affected. This is especially important in cases like Quinacrine, where the evidence base is problematic and societal values between different actors conflict causing a lack of trust and a history of miscommunication.

The role of Quinacrine critics in these processes cannot be underestimated. They are vital partners in seeking accountability for socially disadvantaged women, whose interests are often not adequately protected through conventional checks like individual informed consent forms or expert review boards (Schüklenk 2000). The challenge of such coalitions is to move from adversarial criticism to collaborative, constructive criticism. In essence to be cooperative without being coopted.

Change cannot be negotiated without the opening of genuine institutional opportunities. In India in particular, activists are many times still not recognised as legitimate actors or valued experts within the policy process. They are often consulted too late, when policy documents have already passed the discussion and drafting stage, leaving very little leeway to incorporate their inputs (Contact No. 2). More fundamentally many are wary of participating in government meetings because of past experiences when they were only invited to participate as silent observers and subsequently were publicly misrepresented by government officials as endorsing government policies (Interviewee No. 13, 20).

Obviously the rebuilding of trust and common ground between all the actors involved can only take place incrementally over time. In addition to the constant negotiation and advocacy skills required for such a change, Schneider and Ingram (1993) suggest that in order for such changes to be sustainable the social construction of the role and value of activists in the eyes of policymakers must also be transformed. Only when activists are seen to be powerful and their expertise seen to be valuable will governments more proactively seek to distribute policy gains in their direction.

In a world marked with tremendous inequalities and divergent histories sensitive engagement is filled with controversial decisions and alliances. Some may feel that they do not have the resources to engage in policy change, yet they also regard those who visibly collaborate with government and international agencies with suspicion (Interviewee No. 13). Some of this distrust is valid and is important to protect sceptical objectivity. However, when suspicion reifies stereotypes about actors, this can discount potential avenues and resources for change. As the case of Quinacrine shows, white, male doctors and female philanthropists from the North, as well as female doctors from the South are firm proponents of Quinacrine. Critics who have opposed the ethics of Quinacrine research, span male and female scientists and activists in the North, international technical and donor agencies, as well as Southern researchers and activists. I believe it is critical to go beyond simple dichotomies of activists vs. scientists, public health vs. biomedical perspectives, feminist vs. non-feminist groups, national vs. international boundaries. Conventional stereotypes only serve to replicate hierarchies of power and barriers of distrust. In order to promote a new politics of democracy, constitutive models of policymaking that probe beyond conventional stereotypes to engage in the diversity of positions and perspectives that exist are required.[16]

As mentioned earlier, the EPA turned to external science advisors to restore their technical legitimacy. Nonetheless, rather than an exclusive return to scientific expertise, the EPA sought processes which 'blended the political legitimacy of wide consultation with the scientific legitimacy of peer review' (Jasanoff 1992: 215). By recognising that risk assessment and regulation involves both scientific and political actors it sought mechanisms for both political accountability and scientific credibility. For example in 1980, the EPA joined the automotive industry in establishing a non-profit organisation to fund research on health effects related to automobile emissions. The Health Effects Institute sought to reduce the adversarial conflicts over science by securing joint funding and protected scientific credibility and autonomy through a series of advisory panels and review procedures. Similarly, policy change by the EPA over the regulation of dioxin did not take place until extensive inter-agency consultation, review by the Science Advisory Board and an international meeting bringing together scientists and regulators had taken place (Jasanoff 1992). The literature on environmental health regulation is beyond the scope of this chapter, but I hope my discussion of one part of it will encourage those involved in developing regulatory regimes for contraceptive research to explore the lessons learned from this field.

Although Quinacrine is often trivialised as a marginal issue within contraceptive research, I believe that the dynamics of this case reflect ongoing debates with respect to contraceptive research and use by poor women the world over. Science serves as a powerful mantel for policy due its technical competence and assumed neutrality. My argument is not against the pervasive use and need for the technical specialisation and insight that expertise can provide, rather I argue that deference to expertise should not take place without taking into consideration its own social context. It cannot be immune from analyses about how it achieves its authority, who in society confers it such power and what are the societal consequences of such deference.

Clarke in examining the history of reproductive sciences reminds us how issues of social control and morality have always marked research in the reproductive sciences. Controversy 'does not necessarily mean that that world is doomed to failure. Rather, it means that an analysis of power and politics within the arena is requisite for participants [...] For there is no consensus, merely successful or unsuccessful negotiations among those involved at any historical moment.' (Clarke 1990: 31).

With this in mind I conclude by emphasising that both science and politics play critical roles in the arena of contraceptive research and regulation. Social analysis of the causal beliefs, knowledge construction and policy strategies used by coalitions provide valuable understanding as to their persistence in the face of political adversity, scientific bankruptcy and regulatory illegality. In these instances it is perhaps naïve to expect closure by following strategies that focus on identifying the 'correct' science and excommunicating the 'incorrect' scientists.

It is my belief that it is not just the lack of inconclusive evidence that spurs the controversy around Quinacrine. The divergence in beliefs, values and social inequalities between the participants involved will continue to spur controversy if and when better evidence arrives. It is therefore imperative that the neglect of social and institutional mechanisms that support further dialogue, transparency and democracy between all participants in the world of contraceptive research and regulation be addressed. An essential component of these mechanisms is improved accountability through not just access, but outreach of information for all participants involved. Without access to full information, monitoring and enforcement activities by government, civil society or communities is made that much more difficult. In India, where deteriorating public services and a dominant, unregulated private health sector exist, these mechanisms are vital not only for overseeing controversial contraceptive research, but also in the use of existing, more accepted contraceptives.

Notes

1. This chapter is derived from research originally undertaken for my Masters thesis. The majority of my data was collected during August 1997 and December 1998. I collected a range of documents (journal articles, newspaper stories, reports compiled by organisations, minutes of meetings, study protocols, newsletters, petitions and private correspondence) and undertook open-ended interviews with 23 out of 31 participants contacted (Appendix B). I undertook this exploratory investigation with no formal training in qualitative research. Reflecting on the tensions that I encountered confirmed to me the importance of defining research as a collective process. Explicit ethical guidelines and professional support must counterbalance the power dynamics that can make research as an individual journey inherently problematic. Issues that proved to be challenging include, among others, handling interviewee-interviewer inequalities during data collection, preserving anonymity during analysis, managing data over long periods of time and addressing my own biases.
2. Initial trials using Quinacrine as a slurry in women report no deaths. However, the manufacturer of the drug at that time, Winthrop Pharmaceuticals, reported three deaths following the use of quinacrine as a slurry (Mulay 1997).
3. Now known as Engendering Health.
4. I focus on the main Quinacrine promoters in India, however more than the four mentioned exist.
5. Formerly known as the International Fertility Regulating Programme.
6. Sita Bhateja subsequently disassociated herself from the CHIP Trust.
7. 'Certainly, the relative risk of having a baby in a rural area is much higher than anything quinacrine presents' Dr Tim Black, Director of Marie Stopes International, cited in DiConsiglio 1994.
8. Quinacrine is already available for the treatment of giardia.
9. Kini calculates that there is a 'supply-demand mismatch' of nearly 17 percent, considering that 47 percent is the 'ideal' rate of sterilisation that must be achieved. In order to reach this rate 'an estimated 8 million additional sterilizations per year will be required' (Kini undated).
10. For example, *International Journal of Gynecology & Obstetrics India* has Rohit Bhatt as Editor-in-Chief, Anumpam Agarwal as Excecutive Editor, Ashi Sarin and Pravin Kini as Associate Editors. All are advocates of Quinacrine sterilisations.
11. 'It is most respectfully submitted that almost every drug carries some harmful effects. Well-qualified and highly trained medical scientists conduct clinical trials to make a risk-benefit analysis of each drug' (Jain undated).
12. *The Hindu* reported a 50 percent failure rate based on their interview of Dr Badri Saxena, ICMR.
13. The development of Quinacrine is seen to be especially important to control Muslim communities who are feared to have higher fertility rates (Interviewee No. 3, 12, 20).
14. Critics span a broad range of local, national and international organisations and networks, some are listed here: Saheli, Mahila Jagorti, Manasa Tingala Patrike, AIDWA, Drug Action Forum, HealthWatch India, Women's Global Network on Reproductive Rights (WGNRR), Health Action International (HAI), Committee for Women, Population and the Environment (CWPE),

the Boston Women's Health Collective, International Women's Health Coalition, *Reproductive Health Matters,* Population Council, AVSC, World Health Organisation.
15. Myself included.
16. This is especially important when recognising the limits of more conventional technocratic and linear policy models. In order for these linear models to succeed, they require consensus about the nature of the problem faced, existence of and certainty in knowledge about the problem, clearly identified and relevant disciplines of expertise, a convergence of societal values and a bounded space to express and discuss expert opinion (Jasanoff 1997).

Acknowledgements

This chapter would not be possible without the numerous individuals in India and elsewhere who generously shared their time, information and opinions about Quinacrine with me. Equally important are my friends and family members without whom I would not have been able to undertake this research. In addition to those mentioned in my thesis, I thank Meena Shivdas, Peter Salama and Ian Urbina for their comments and Maya Unnithan for her support. Most importantly I am indebted to Abha Sur for her inspiring friendship.

References

AVSC. 1993. 'AVSC Technical Statement: Quinacrine Pellets for Nonsurgical Female Sterilization'. Washington, D.C.: Engender Health

Benagiano, G. 1994. 'Letters to the Editor: Quinacrine family planning method'. *The Lancet,* 343: 1425

Canadian Women's Committee on Population and Development. 1996. *Bill of Rights for contraceptive research, development and use.* Ottawa: Canadian Women's Committee on Population and Development

Carignan, C. S., Rogow, D. and Pollack, A. E. 1994. 'The Quinacrine Method of Nonsurgical Sterilization: Report of Experts Meeting'. AVSC Working Paper. 6. Washington, D.C.: Engender Health

Clarke, A. E. 1990. 'Controversy and the Development of Reproductive Sciences'. *Social Problems,* 37, 18–37

Dasgupta, R. 2001. 'Another Chapter Begins: The Ongoing Saga of Quinacrine Sterilisation in India'. *Political Environments, A Publication on Women, Population and the Environment,* Winter/Spring, 46–48

DiConsiglio, J. 1994. 'Risks and Rewards: Family Planners Weigh Quinacrine'. *Family Planning World.* January/February: 1, 20

Gandhi, N. and Shah, N. 1992. *The issues at stake: theory and practice in the contemporary women's movement in India.* New Delhi: Kali for Women

Germain A. undated. 'Are we speaking the same language? Women's health advocates and scientists talk about contraceptive technology'. Internet: www.iwhc.org/febc1.html

Gieryn, T. 1999. *Cultural Boundaries of Science: Credibility on the Line.* Chicago: Chicago University Press

Haas, P. M. 1992. 'Introduction: epistemic communities and international policy coordination'. *International Organization,* 46: 1, 1–35

Harding, S. 1992. 'After the Neutrality Ideal: Science, Politics, and Strong objectivity', in M. Jacobs, ed. *The Politics of Western Science*. Palo Alto: Stanford University Press, 80–101

Hardon, A. P. 1992. 'The needs of women versus the interests of family planning personnel, policy-makers and researchers: Conflicting views on safety and acceptability of contraceptives'. *Social Science and Medicine*, 35: 6, 753–66

Hartmann, B. 2001. 'Quinacrine Update: FDA Grants Approval for Human Trial'. *Political Environments, A Publication on Women, Population and the Environment*. Winter/Spring: 45

Haskell, T. L. 1990. 'Objectivity is not Neutrality: Rhetoric vs. Practice in Peter Novick's'. *That Noble Dream. History and Theory*, 29: 132

Hieu, D. T., Tan, T. T., Tan, D. N., Nguyet, P. T., Than, P. and Vinh, D. Q. 1993. '31,781 cases of non-surgical female sterilisation with quinacrine pellets in Vietnam'. *The Lancet*, 342, 870–1

Jain 1997. 'Letter to Editor, Press Trust of India'. *Quinacrine Sterilisation Newsletter*, 2: 1, 5–6

—— undated. Notes in Response to AIDWA/JNU Case

Jasanoff, S. 1992. 'Science, Politics, and the Renegotiation of Expertise at EPA'. *OSIRIS*. 2nd Series, 7, 195–217

—— 1997. 'Civilization and Madness: The Great BSE Scare of 1996'. *Public Understanding of Science*, 6, 221–32

—— 1998. 'The Political Science of Risk Perception'. *Reliability Engineering and System Safety*, 59, 91–9

Keck, M. E. and Sikkink, K. 1998. *Activists Beyond Borders: Transnational Advocacy Networks in International Politics*. Ithaca: Cornell University Press

Kessel, E. 1994. 'Commentary'. *The Lancet*, 344, 698–700

—— 1996. '100,000 quinacrine sterilizations'. *Advances in Contraception*, 12, 69–76

—— 1997. 'Quinacrine pellet sterilisation update'. *International Journal of Gynecology & Obstetrics India*,1: 1, 69–74

Kini, P. 1996. *Quinacrine Sterilization Newsletter*,1: 1, 2

—— Undated. 'Innovative Method of Defusing the Population Bomb'. Unpublished article

Kini, P., Bhateja, S. and Rajgopal. Undated. 'An overview of quinacrine non-surgical female sterilisation and report of an initial trial at Bangalore and Mangalore in Karnataka, India to popularize Q method of sterilization'. Unpublished article

Lancet, The. 1994. 'Editorial: Death of a study: WHO, what and why'. *The Lancet*, 343, 987–88

Latour, B. 1987. *Science in Action*. Cambridge: Harvard University Press

Latour, B. and Woolgar, S. 1986. *Laboratory Life*. Princeton: Princeton University Press

Mendonca, S. 1994. 'The Q method: A new method of female contraception is entangled in bureaucratic web'. *Sunday*. September/October: 82

Mulay, S. 1997. 'Quinacrine Sterilisation: Unethical Trials'. *Economic and Political Weekly*. April 26, 877–79

Mumford, S. 1994. 'Quinacrine Pellet Method for Nonsurgical Female Sterilisation Progress Report. September 30'. 1–6. Center for Research on Population and Security

NCESSRH. 2000. *Ethical Guidelines for Social Science Research in Health*. Mumbai: Cehat

Network. 1995. 'FHI Quinacrine Studies'. *Network.* September, (27). Family Health International

Oudshoorn, N. 1994. *Beyond the natural body: An archeology of sex hormones.* London: Routledge

Quinacrine Sterilization Newsletter. 1997. 2: 1, 2

Pollack, A. E. and Carignan, C. S. 1993. 'The Use of Quinacrine Pellets for Non-surgical Female Sterilisation'. *Reproductive Health Matters,* 2, 119–22

Rao, M. 1997. 'Quinacrine "Trials" and the National Security Questions in the USA'. *Women's Global Network for Reproductive Rights,* 1: 7

Ravindran, S. and Berer, M. 1994. 'Contraceptive Safety and Effectiveness: Re-Evaluating Women's Needs and Professional Criteria'. *Reproductive Health Matters,* 3, 6–11

Rogers, W. 1979. 'A hard look at Vermont Yankee: Environmental Law Under Close Scrutiny'. *Georgetown Law Journal,* 67, 699–727

Saheli. 1997. *Quinacrine: The Sordid Story of Chemical Sterilisations of Women.* New Delhi: Saheli

Schneider, A. and Ingram, H. 1993. 'Social Construction of Target Populations: Implications for Politics and Policy'. *American Political Science Review,* 87: 2, 334–47

Schüklenk, U. 2000. 'Protecting the vulnerable: testing times for clinical research ethics'. *Social Science and Medicine,* 51, 969–77

Sen, G. 1995. 'Rights and Reproductive Technologies', in M. Schuler, ed. *From Basic Needs to Basic Rights.* Institute for Women, Law and Development. Washington, D.C: Institute for Women, Law and Develoment, 391–400

Simmons, R., Hall, P., Diaz, M., Fajans, P. and Satia, J. 1997. 'The Strategic Approach to Contraceptive Introduction'. *Studies in Family Planning,* 28: 2, 79–94

Spicehandler, J. and Simmons, R. 1994. *Contraceptive Introduction Reconsidered: A Review and Conceptual Framework.* WHO/HRP/ITT/94.1. Geneva: World Health Organisation

The Hindu. 1997. Unsafe contraceptive method prescribed. April 2

Wieringa, N. and Hardon, A. 1993. Women's Health Action Foundation. Public Address

Women's Global Network for Reproductive Rights 1996. 'Call for a Reorientation of Contraceptive Research'. *Newsletter,* April–June (54): 15

Zipper, J., Stachetti, E. and Medel. M. 1970. 'Human fertility control by transvaginal application of quinacrine on the fallopian tube'. *Fertility & Sterility,* 21, 581–89

—— 1975. 'Transvaginal chemical sterilisation: clinical use of quinacrine plus potentiating adjuvants'. *Contraception,* 12, 11–21

Appendix A: List of Organisations Mentioned

AIDWA	All India Democratic Women's Association
AVSC	Association for Voluntary Surgical Contraception now known as Engendering Health
BHWC	Boston Women's Health Book Collective
CRPS	Center for Research on Population and Security
CWPE	Committee for Women, Population and the Environment
CHIP Trust	Contraceptive and Health Innovations Project
DCI	Drug Controller of India
DTAB	Drug Technical Advisory Board
FHI	Family Health International
HAI	Health Action International
	Health Watch
HRP	Human Reproduction Programme of WHO
ICMR	Indian Council for Medical Research
IDRC	International Development and Research Center
IFFH	International Federation of Family Health
IWHC	International Women's Health Coalition
JNU	Centre for Social Medicine and Community Health at the Jawaharlal Nehru University
UNFPA	United Nations Population Fund
USAID	United States Agency for International Development
WGNRR	Women's Global Network on Reproductive Rights
WHO	World Health Organization

Appendix B: Lists of Respondents

Contacted

Parthas Das Gupta	Drug Controller of India	Aug 97	
Saraswati Ganipathi	Belaku Trust	Aug 97	
Imrana Quadeer	Jawarhalal Nehru University	Aug 97	
V. Ramalingaswami	Former director of ICMR	Aug 97	
Badri Saxena	Associate Director of ICMR	Aug 97	
Mira Shiva	Voluntary Health Association of India	Aug 97	Dec 98
Nalini Viswanathan	CWPE	Nov 98	
Sita Bhateja	CHIP Trust	Dec 98	
Viswanath	Marie Stopes Clinic	Dec 98	

Interviewed

Anupam Agarwal	Jain Clinic	Dec 98	
Indu Agnihotri	Centre for Women and Development Studies	Dec 98	
Stena Almorth	Population Council	Dec 98	
Vineeta Bal	Indian Institute of Immunology	Dec 98	
Nilanjana Biswas	Manasa Tingala Patrike	Dec 98	
Nandita Gandhi	Akshara	Aug 97	
Krishna Jaffa	Jain Clinic	Dec 98	
Brinda Karat	All India Democratic Women's Alliance, New Delhi	Aug 97	Dec 98
Pravin Kini	CHIP Trust	Aug 97	
Parvathi Menon	Frontline	Aug 97	
Stephen Mumford	Center for Research on Population and Security	Nov 98	
Shree Mulay	McGill University	Nov 97	
Lakshmi Murthy	Saheli	Aug 97	Dec 98
Padma Prakash	Economic and Political Weekly	Aug 97	
Mohan Rao	Jawarhalal Nehru University	Aug 97	Dec 98
Prakash Rao	Medical Clinic	Aug 97	
Sundari Ravindran	Reproductive Health Matters	Aug 97	Dec 98
Sarojini	Magic Lantern Foundation	Aug 97	Dec 98
Amit Sen Gupta	All Indian Drug Action Network	Aug 97	
Gita Sen	Indian Institute for Management, Bangalore	Aug 97	Dec 98
Kirti Singh	Lawyer	Dec 98	
Sushma Varma	Mahila Jagorti	Dec 98	
Vimla	All India Democratic Women's Alliance, Bangalore	Aug 97	

CHAPTER 8

'SHE HAS A TENDER BODY':
POSTPARTUM MORBIDITY AND CARE DURING
BANANTHANA IN RURAL SOUTH INDIA

*Asha Kilaru, Zoe Matthews, Jayashree Ramakrishna,
Shanti Mahendra and Saraswathy Ganapathy*

'*She has a tender body. If we do not do a strict bananthana, she will be weak in later life, she should become like the tip of a mantani leaf – thin, tender, fresh and supple.*'

Introduction

Postpartum deaths in developing countries are more common than those occurring during pregnancy and delivery (Li et al. 1996). Although most postpartum deaths occur in the first twenty-four hours after delivery, a significant minority occur after the first day, when, in the context of home deliveries in rural areas, trained health care providers are no longer assisting the recently delivered mother. In a study of deaths in rural Bangladesh, 62 percent of all non-abortion maternal deaths occurred in the postpartum period, and of these 56 percent occurred between two and ninety days following delivery (Fauveau et al. 1988). In these circumstances, the understanding of danger signs and the subsequent healthcare-seeking behaviour of women and their families becomes a crucial factor in maternal survival and wellbeing. The development of effective strategies to improve postpartum health requires an improved awareness of sociocultural influences on care-seeking which this paper seeks to provide in the South Indian context.

The World Health Organization (WHO 1998) provides recommendations for the management of postpartum health problems. Unlike prenatal and intrapartum care, where standards are available to healthcare providers and communities, postpartum care protocols are usually lacking in developing countries. The WHO, bringing together the very few studies available that concentrate on this period, recommends that

> postpartum care must be a collaboration between parents, families, caregivers trained or traditional, health professionals, health planners, health care administrators and other related sectors such as community groups, policy makers and politicians. (WHO 1998: 49)

To ensure the acceptance of new guidelines for postpartum care, it is important to see how they can be adapted to traditional views. By virtue of its prospective design, this study is uniquely placed to provide insights into care-seeking behaviour as pregnancy and postpartum periods evolve. The extent of the gap between cultural and biomedical models of good postpartum health has been explored. Furthermore, an examination of the actual provision of services in the region is also undertaken.

The study setting is Karnataka, where the rural population is still underserved in terms of facilities, and maternal mortality and morbidity is high (Bhatia 1993, 1995). However, in the Indian context, the state does maintain a progressive profile in terms of health programmes, with recent survey data confirming that 51 percent of deliveries take place in institutions as compared with a national average of 34 percent (IIPS 2000). Despite this relatively favourable aggregate position, the state shows marked regional diversity. Rural areas are disadvantaged compared with the urban areas with only an estimated 26 percent of rural births taking place in institutions as compared with 67 percent in urban Karnataka (Macro International 1993). In summary, the study location provides a background of remote services in a resource poor rural area, but with the potential and infrastructure for further improvements in the near future.

In this part of southern Karnataka, the postpartum period is culturally well recognised and defined. There are specific terms for a woman in the postpartum period (*bananthi*) as well as for the period itself (*bananthana*). At this time the mother and child are thought to be physically and psychologically vulnerable and easy prey to illnesses of natural and supernatural origin. Delivery is considered a polluting process from which all participants must be ritually cleansed. The newly delivered mother is restricted in terms of mobility and diet, for the sake of her own and her baby's well-being. Thus, many cultural factors play an important part in the chain of decision making that might eventually lead to contact with medical services.

Self-Reported Morbidity and Health Care Seeking

Previous studies have suggested that the level of maternal morbidity in developing countries is high (Koblinsky et al. 1992). However, estimating the burden of disease or the prevalence of various categories of ill health is fraught with a well-documented range of problems (Ronsmans 1996,

Fortney and Smith 2000). Evaluation of postpartum morbidity is rare and literature surrounding the post-delivery period is still noticeably sparse, despite indications that this is a period of significant ill health and mortality worldwide. For example, in a community survey of 207 women in rural Egypt, 60 percent of respondents reported a postnatal problem (El-Mouelhy et al. 1994). Most common amongst these were discharge and fever, followed by depression and bleeding. A larger study conducted in Karnataka in 1993, based on 3600 recent mothers, found that 33 percent of cases reported symptoms of current reproductive morbidity (Bhatia and Cleland 1995). Hospital studies have also estimated prevalence (Prual et al. 1998), but these suffer selection bias both in terms of increased likelihood of problems as well as those who can afford care.

Most studies measuring postpartum morbidity rely on self-report, which is unreliable when compared with medical assessments (Stewart et al. 1996, Ronsmans et al. 1997). Low prevalence conditions, such as many of the individual postpartum morbidities, are particularly prone to overestimation. Further, the quality of self-reports varies considerably according to levels of education, socioeconomic status and access to care. Despite these problems, self-reports in response to well designed and carefully worded interviews may be the only way to collect morbidity information where most births take place outside of institutions. Self-reports are critical for understanding how women interact with local health care providers (Fortney and Smith 2000) and as a marker for potential demand of services. In their conceptual framework of maternal morbidity, Murray and Chen (1992) assert that self-perceived morbidity measures important aspects of an illness that are fundamentally different from observed morbidity, as it is a function of disease pathology, cultural context and socioeconomic factors.

Cultural context has an even more important role to play in the understanding of health seeking patterns. This is particularly true of the postpartum period, where a woman's health is more separate from that of the child's than it has been during pregnancy. Furthermore, the new status of 'mother' is bound by cultural expectation and tradition in all societies. In a qualitative study of beliefs and practices regarding postpartum morbidity in rural Bangladesh, postpartum seclusion, restrictions on diet and mobility and the concept of delivery as polluting are seen as key factors in the understanding of this socially prescribed period (Goodburn et al. 1995). Such factors are similar to those in Karnataka, and may apply in modified form to a substantial proportion of mothers in the world. More quantitative studies such as the earlier Egyptian investigation into pregnancy-related morbidity, have been able to look at the differential reporting of problems between older and younger women, and between primigravidas and higher parity mothers (El Mouelhy et al. 1994).

Study Design, Methods and Setting

The study described in this paper, funded by the WHO, aimed to identify the sociocultural determinants of obstetric morbidity in rural Karnataka by establishing the type and extent of obstetric morbidities and explore health-seeking behaviours and factors affecting service uptake during the obstetric period. A related objective was to gather information on traditional beliefs and practices and to explore the nature of health services provided in the rural context. Both quantitative and qualitative methods were used, and a prospective research design was employed in order to overcome the recall biases inherent in retrospective studies.

The study was carried out in eleven villages covering a population of approximately 25,000 surrounding a town where the taluk headquarters is located (many taluks comprise one district), located about sixty kilometers from Bangalore city. The closest of the study villages is about eight kilometers from the taluk headquarters town and the furthest about twenty-five kilometers. The study villages were selected randomly from the villages in the taluk, with the later addition of a small town and a tribal village, in order to capture health-seeking behaviour in a range of rural settings.

All women in these villages who were already pregnant at the time that the study began in August 1996 or who became pregnant during the study period were enrolled until the required sample of five hundred cases was reached. The entire process of the panel survey was completed within twenty-five months of the start date.

Respondents were visited five times during the study. These five visits consisted of an initial questionnaire, followed by two successive antenatal questionnaires, a post-delivery questionnaire administered three to five days after delivery, and finally, the postpartum questionnaire. This last questionnaire, which forms the basis of the analysis presented in this paper, was carried out three to five months after delivery.

Morbidity questions were treated with particular care, the women's perceptions of ill-health being elicited without prompts or preset categories, and only subsequently making recourse to a more structured list of possible morbidities. Clearly the normal biases of prospective studies could not be avoided including a raised awareness of obstetric issues as the study progressed. However, an examination of results by duration of study does not reveal any significant increase in morbidity, or changes in health seeking behaviour.

As in many other surveys in rural areas, respondents were quite happy to participate in the study. There was, however, a high rate of drop-out during the course of the study. Of an initial sample of 535 women who completed the initial questionnaire, only 366 were re-interviewed during their bananthana (postpartum) period. The women who dropped out of the study were all, without exception, lost to follow up because they returned to their natal village to deliver, a common custom in this area. On examining the characteristics of the drop-out population, the parity, age and socioeconomic distribution was found to be very similar to that of the study group.

Karnataka has a profile typical of South India, with early marriage for women (the median age at marriage is approximately 16.8 years for women) and consanguineous marriages are common. Recent rapid fertility declines at all ages have brought the Total Fertility Rate in Karnataka to 2.25 for rural residents, with a predominance of short birth intervals, almost half of which are less than two years in duration (IIPS 2000). Use of contraception is high within the state with female sterilisation being the preferred method, as in most other Indian states (Macro International 1993). Literacy levels in the state are just a little higher than the national average of 52 percent, with male rates exceeding female rates by over 30 percent (Bhatia and Cleland 1995).

Characteristics of the Study Population

As seen in Figure 8.1, almost three-quarters of the 366 sampled women in the study villages were between 18 and 24 years old at the start of their pregnancies and 14 percent were under 18. Just over half of the women had attended school for some time, and almost half of these had reached class 9 or 10 (10 being the equivalent of high school).

Survey data on occupations showed that all women were engaged in household work, but only 2 percent as salaried work or trading as a primary occupation and 8 percent as waged agricultural work as a primary occupation. Those who carried out agricultural wage labour were among the poorest in the sample. Most belonged to households which have small landholdings, and one quarter of the households were landless.

Age at *prastha*, when sexual relations with the husband commences, is very early in this population. Thirty six percent of the women reported that they were under 16 at that time and 28 percent reported that they were 16 or 17 years old. More than a third of the women were pregnant for the first time, and 36 percent for the second.

The sample is fairly typical of rural south Karnataka in terms of caste distribution, with the Other Backward Caste (OBC) comprising the majority of the respondents, followed by a significant minority of 29 percent Scheduled Castes and Scheduled Tribes (SC/ST) and finally the Lambani tribe within the SC/ST who live in a separate village.

A geographical categorisation was created to locate the villages within areas that had access to the same Auxilliary Nurse Midwife (ANM) or health subcentre. Group 1 consists of the four villages on the western side of the taluk that are served by the same subcentre. Group 2 consists of a more disparate group of five villages in the centre north of the area; a subcentre located centrally in the group serves these villages. After this categorisation, there remained the large village, situated in the southernmost part of the study area, and served by a number of health

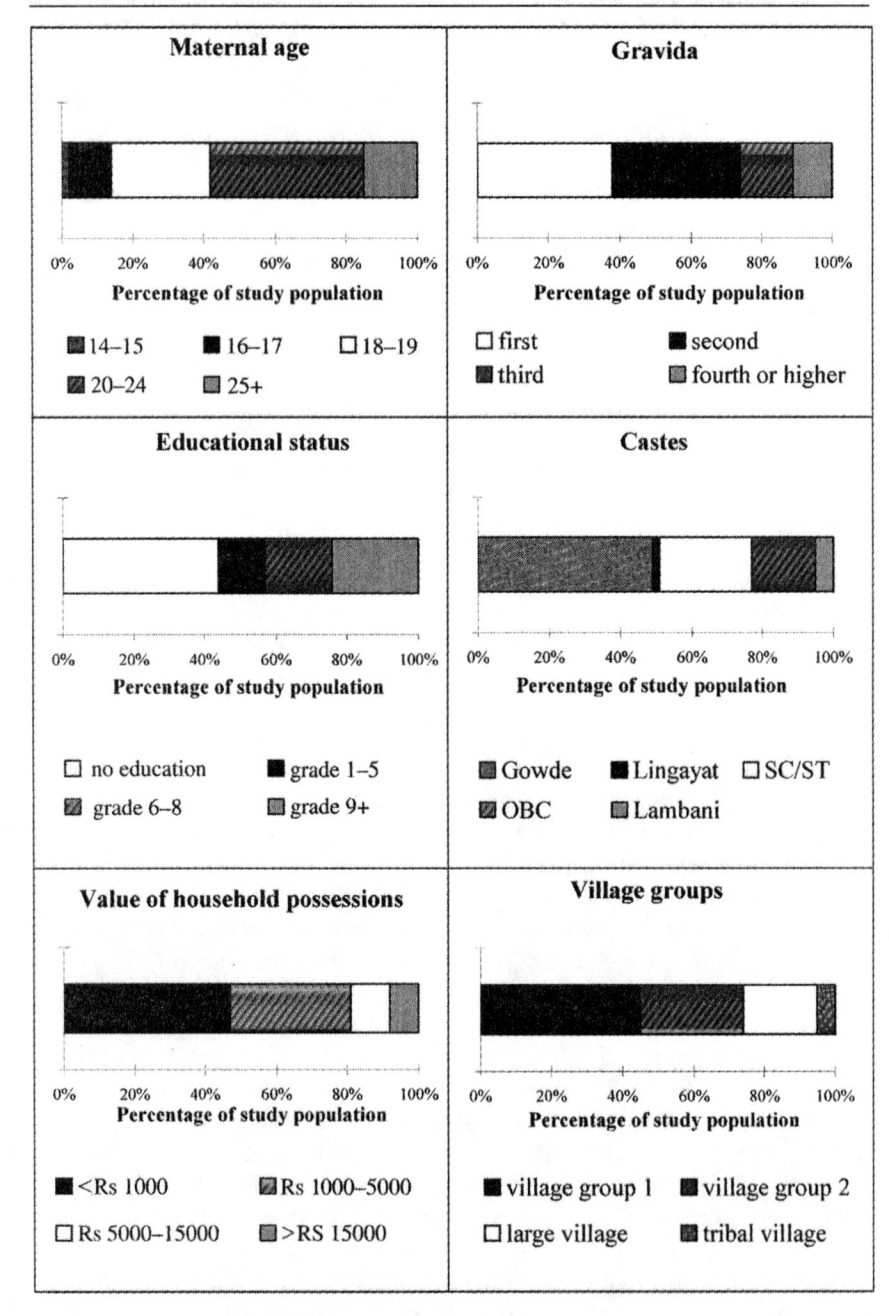

Figure 8.1. *Characteristics of the Study Population*

providers. Lastly the tribal village, consisting of Lambani tribespeople only, was considered as a separate category.

Traditional Care during Bananthana

In this part of southern Karnataka, the postpartum period is culturally well recognised and defined. There are specific terms for delivery (*herige*), and womb (*karalu*; also used to refer to the intestines). There is no equivalent term to bananthanal in English; the concept of 'confinement' may be the nearest approximation. The aim of the traditional bananthana, according to key informants in the study area is threefold:

- To promote the baby's and mother's health
- To prevent illnesses in the baby and the mother
- To ensure that the mother does not have any long term disability or infirmity that will show up later as she gets older

In general, it is seen as a very special period in a woman's life, where a strict traditional health regimen applies, not only for the well-being of the child, but also for the future well-being of the mother. The following quote from an older female relative of a bananthi exemplifies the positive attitude of care expressed for women in the postpartum period: 'She has a 'tender body' (*hasi mayi*). If we do not do a strict bananthana, she will be weak in later life'.

From the qualitative interviews it was found that local people believe pregnancy is a time when 'dirty' or 'bad fluids' are accumulated in the body. A typical statement is that: 'All the bad water should be drained from the body'. 'Hot' foods (those foods that produce bodily 'heat' under the ayurvedic system) are seen as essential for eliminating these fluids. Ideally within a three month period a bananthi has to lose these fluids and become 'half' her size, according to local custom. Furthermore, the local perception is that a recently delivered woman's 'karalu' (stomach or womb) is 'hasi' (tender) as well as inflamed and loose. So she should therefore eat only small quantities of bland food. 'Cold' foods (those which reduce bodily heat under the ayurvedic system) are avoided. Moreover, the food that a mother eats is seen to affect her breastmilk that in turn affects the baby's health. If a lactating mother eats 'cold' or 'gas producing' foods the baby is believed to be prone to problems such as colic or diarrhoea.

During this period, the mother may be affected by 'drishti' (evil eye), or possessed by spirits and therefore afflicted with various illnesses. It is believed that 'cold' must be prevented from entering the body through the birth canal. It is believed that keeping the legs crossed can prevent this. The rituals and practices of bananthana are organised to afford maximum protection to the mother and child. Bananthana extends from the time of delivery up to a varying endpoint, commonly three months, but this sometimes extends up to five or seven months, and is often determined by the family's socioeconomic status.

Delivery is considered a ritually polluting process. Bananthana is divided into discrete periods that mark reductions in the degree of pollution. These periods are marked by changes in diet, in mobility, in health promotive and preventative measures. It can be roughly divided into four periods; the transition from one period to the next is marked by rituals and gradual resumption of diet, activity and mobility. The ideal periods are outlined in Table 8.1.

Table 8.1. *Traditional Bananthana Practices by Duration of Postpartum Period*

Time period	Practices
delivery to day three	• mother and child are confined to a warm, dark room • mother and child are kept warm • access to mother and child is limited to care givers and immediate family • mother is given hot, freshly prepared food like rava ganji (cream of wheat porridge) with jaggery (molasses) twice per day, also coffee with jaggery • fluid intake is restricted, hot water only after meals • betel leaf and betel nut with lime are given to mother • pre-lacteal feeds given to baby including castor oil, sugar water and diluted cow's milk, also wetnursing • initiation of breastfeeding • sponge bath or body bath given for both infant and mother • umbilical cord care: baby is branded around the navel and other points • mother lies down most of the time and does not do any work • a binder is tied around the abdomen to support the back and to keep intestines and uterus in place
day three to day eleven	• mother given bland rice with salt • fluid restriction continued • head bath given for mother, neem and turmeric paste smeared before bath, hair dried with samrani (resin) fumes • some initiate breastfeeding after head bath (it is a common belief that the milk 'descends' after the head bath) • mother continues to be protected from 'cold' entering the body • no housework done by mother
day eleven to day forty	• head bath given for mother and child • alcohol (usually brandy or arrack, 1–2 pegs) given to mother after head bath to ward off 'cold' • ritual pollution period ends – house cleaned and new pots brought • mother is given freshly prepared rice and warm 'hot' foods including chicken, goat meat and dried prawns with extra 'hot' foods such as pepper, garlic and cumin • mother given very few vegetables; those that are given are almost all greens, gourds and foods that produce flatulence: no 'cold' vegetables • sesame seed oil (utcche yellu) given to mother • ideally no work done by mother, and no contact with cold water
day forty to three/four months postpartum	• after visiting the temple, mother and child can leave the house for essential visits • mother may start eating 'ragi'(millet) balls • mothers from affluent households with social support continue to rest

Boys are considered constitutionally weaker than girls are, and thus the mother is not supposed to relax the restrictions. Also, with a boy baby, the mother must eat less food, if she eats too much, it is feared that the baby will not be able to digest the milk and will vomit.

Most young women, regardless of educational level, follow the advice of their elders, particularly their mothers for fear of jeopardising their baby's health. The following are some examples from postpartum narratives of bananthis, part of the qualitative data collected:

- Pavithra, who has several years of education and has given birth to her first baby, a boy, was bored during her bananthana period. She wanted to drink water but was restricted. She knew that if she kept her legs apart, then she risked her 'karalu' coming down, so that she always tried to stay lying down while keeping her legs crossed so that the birth canal could become narrow. She also accepted that the bananthi must always be strict; she must eat very well only once per day and drink as much water as she wants at that time, cover herself and sleep. In this way she hoped to bring out all of the bad water in her body through perspiration. Pavithra followed these traditions because she wanted the child to be healthy.
- Sunitha, who delivered her first baby said that a bananthi should rest and lie down all of the time. She should not sit for any length of time because, although she may not feel the pain now – she is sure to feel pain in fifty years time.
- Rathnamma gave birth to a son after two daughters. Her mother insisted that she follow a strict bananthana for this precious male child. Only after her sterilisation was she given enough food to fill her stomach.

Though the confinement and restricted diet is irksome to many women, some seemed to have enjoyed their bananthana. They may follow their doctor's advice and therefore not follow certain customs, but if this happens, they often feel that they missed out on something important.

Routine Postpartum Contact with Modern Medical Providers

Local cultural norms do not include any specific seeking out of healthcare services. The woman who conducts the delivery, whether ANM, *dai* or lay woman, usually gives advice about practices immediately following the delivery. The newly delivered mother generally has little input. Tradition, as interpreted by the older women conducting the bananthana, lays down the appropriate rituals and practices to be carried out. These practices are considered as completely adequate and beneficial for the continued good health of mother and child, and there is therefore no perceived need for routine medical postpartum checkups. Although the majority of women in the study population seek antenatal care, there is very little commensurate seeking of care in the postpartum period.

In the postpartum period, it can be difficult to distinguish routine checkups from contacts with health care personnel for other reasons, such as baby health checks and advice about contraception. Furthermore, the idea of seeking care must be treated with caution, as routine visits are often made by healthcare personnel (especially the ANM), which implies that the care is not necessarily sought by the woman. Postpartum healthcare contact, in this study, was therefore defined as any contact, whether sought or supplied, for any of the following reasons:

- routine or problem check up for the woman
- routine check, immunisation or problem check for the child
- advice or actual intervention for contraception

Many combinations of the above reasons were given for each postnatal contact recorded, such that a routine contact can also be a contact that contained other elements from the above list.

Although from Table 8.2, only 27 percent of women had no postpartum contact whatsoever, many of these contacts with health providers consisted only of advice from medical personnel regarding sterilisation (tubectomy) or other contraceptive measures. The second column of Table 8.2, which does not include such contacts, more accurately reflects the low rate of postnatal care, most women seeing a provider only once, or not at all. Furthermore, even those contacts that were not directly related to tubectomy advice had a bias towards care for the child, rather than the mother.

Examination of provider contacts should differentiate between routine contacts and those made in response to a problem. While the modern medical healthcare system provides for repeated postnatal visits by the ANM, this does not take place very frequently. There are some areas where the ANM never visits. Of the 366 women in the study, 150 (40.9 percent) had some kind of routine care contact in the forty two days after delivery, whether this was in combination with problem care, or

Table 8.2. *Number of Providers Contacted for any Postnatal Care with and without Tubectomy-Related Contacts*

Health care contact	Number of women with any postnatal care contacts (including tubectomy contacts)	Number of women with any postnatal care contacts (excluding tubectomy contacts)
1 provider contact	137 (37%)	169 (46%)
2 provider contacts	120 (33%)	43 (11.5%)
3 provider contacts	12 (3%)	1 (0.5%)
No contacts with any health care provider	97 (27%)	153 (42%)
Total	**366 (100%)**	**366 (100%)**

Source: Prospective maternity survey.
Note: Care contacts are within 42 days after delivery.

including immunisation and tubectomy advice. A smaller proportion (10.4 percent totalling thirty-eight cases) contacted health providers for a problem (this includes for non-obstetric problems). It is mainly ANMs who provide routine care in the locality. This represents 98 percent of the routine care that was given. In contrast, the contacts made in response to a problem were spread over a variety of providers.

Neither the bananthi nor her husband plays a major role in making decisions regarding her care. Seeking care during bananthana is influenced by many factors, including the family's financial situation, the gender of the child, previous experience with healthcare providers, and a host of contextual and situational factors.

Local perceptions of quality of care are an important factor in healthcare seeking behaviour in this area. Women from the sample reported a great variety of responses when asked about the quality of local health care services. This was mainly dependent on their assessment of the local ANM, or whether their local ANM ever visited. Many respondents criticised the lack of facilities and personnel in the Primary Health Center (PHC). There were comments about the cleanliness of the labour room, and the level of care during pregnancy and delivery. There were no comments or views on postpartum care as such care was not expected. A 'good touch' was also an important characteristic of an effective health provider. One respondent; Shivratnamma said that: 'I go to a doctor's shop in a nearby village. His *kai guna* [literally 'hand quality'] is very good and he attends to us at any time of day. When he treats, we are cured immediately.'

A logistic regression model that set any routine postnatal contact as an outcome variable confirmed the nature of postpartum visiting patterns. The results, (Table 8.3) suggest that the women in village group 2 have three times the odds of a routine postnatal contact as compared with other locations. This is mainly a result of conscientious postnatal visiting by the group 2 ANM, rather than healthcare seeking on the part of newly delivered women. The tribal Lambani women are clearly underserved compared with other women, and those with very high levels of education are less likely to have a postnatal contact. The latter result seems surprising, but is probably due to the highly educated being able to afford private problem care, should the need arise. Finally, the effect of age is caused by the routine contacts that are also combined with tubectomy contacts, where older women are much more likely to be involved.

Table 8.4 shows that many recommended aspects of postpartum care such as the checking of blood pressure were often omitted. The most common procedures reported were the prescribing of iron and folic acid tablets and advice on contraception.

Table 8.3. *Results of a Logistic Regression to find Correlates of Routine Postnatal Contact*

Factor	Level	Parameter estimate (SE)	Odds ratio	N
Location	Large village	reference	1.00	76
	Village group 1	0.47 (0.34)	1.60	163
	Village group 2 **	1.15 (0.35)	3.14	108
Caste	Other Backward Classes	reference	1.00	47
	Scheduled Caste/Tribe*	−1.49 (0.41)	0.23	96
	Lingayat	−0.64 (0.60)	0.53	21
	Lambani *	−2.35 (1.08)	0.10	19
	Gowda	−0.29 (0.40)	0.75	183
Age	Less than 20	reference	1.00	150
	20+ **	0.66 (0.27)	1.94	216
Education	None, or less than Grade 9	reference	1.00	277
	Grade 9+*	−0.53 (0.28)	0.59	89
Constant		−1.17 (0.39)		366

Source of data: Prospective maternity survey.
Notes:
1. Parsimonious models are shown above, 20 candidate variables were tried in the models, and eliminated by backward substitution. Candidate variables were as follows: location, caste, age, education, gravida, possessions score, occupation, problems in previous pregnancy or postpartum periods, land and livestock ownership, family structure.
2. Key to significance levels as follows: * Significant at 5% level, **Significant at 1% – these denote significant differences from the reference category in each case.
3. Note that routine postnatal contact can also be in combination with other forms of contact.

Table 8.4. *Content of Routine Postpartum Care by Type of Health Care Provider*

	ANM or Lady Health Visitor	Government Doctor	Private Doctor provider	By any health care
Abdomen palpated	1 (0.69%)	0 (0.0%)	0 (0.0%)	1 (0.66%)
Iron and folic acid prescribed	14 (9.5%)	0 (0.0%)	1 (100.0%)	15 (10.0%)
Blood pressure	2 (1.4%)	1 (50.0%)	1 (100.0%)	4 (2.7%)
Weight taken	1 (0.7%)	1 (50.0%)	1 (100.0%)	3 (2.0%)
Blood taken	3 (2.1%)	1 (50.0%)	0 (0.0%)	4 (2.7%)
Urinalysis	2 (1.4%)	1 (50.0%)	0 (0.0%)	3 (2.0%)
Vaginal examination	1 (0.7%)	0 (0.0%)	0 (0.0%)	1 (0.7%)
Advice on diet	8 (5.6%)	1 (50.0%)	0 (0.0%)	9 (6.0%)
Advice on breast-feeding	2 (1.4%)	0 (0.0%)	0 (0.0%)	2 (1.3%)
Advice on signs of problems	2 (1.4%)	0 (0.0%)	0 (0.0%)	2 (1.3%)
Advice on contraception	44 (30%)	0 (0.0%)	0 (0.0%)	44 (29.3%)
Total no of women who made contact	144	2	1	150*

Source: Prospective maternity survey.
Notes:
1. Care contacts are within 42 days after delivery
2. *Three additional persons saw two health care providers for routine care, these cases are only included in the final column

Postpartum Problems and Care-Seeking

Clearly there exists a widespread belief that bananthis only need care if a problem occurs. However, from the postpartum questionnaires it was apparent that the occurrence of problems was not rare: 30 percent reported morbidity in the three months following delivery and 15 percent within forty-two days of delivery (this excludes anaemia, non-obstetric problems and fever that lasted less than three days). Other studies have also shown significant morbidity beyond the forty-two-day convention (Defo 1997). Although most heavy bleeding cases occurred within the forty-two-day period, some of the major morbidities such as abdominal pain and abnormal discharge show substantial increase beyond that time. Figure 8.2 shows the distribution of women who reported morbidities for the forty-two-day period.

In the forty-two days after delivery, only half of the women who reported heavy bleeding sought care for their condition largely from the public healthcare system. In contrast, almost all those with fever resorted to medical care. Three additional women mentioned high blood pressure, but clearly only those women whose blood pressure had been measured would be alerted to this. The same is true for anaemia. Three additional women reported malaria or jaundice in the postpartum period, and eighteen reported other non-obstetric problems.

It is believed that too much 'heat' in the post-delivery diet causes excessive bleeding. However, the perception of 'excessive' or 'normal' bleeding is clearly difficult to gauge. Women in the area estimate the amount of bleeding by the number of times that they need to change their

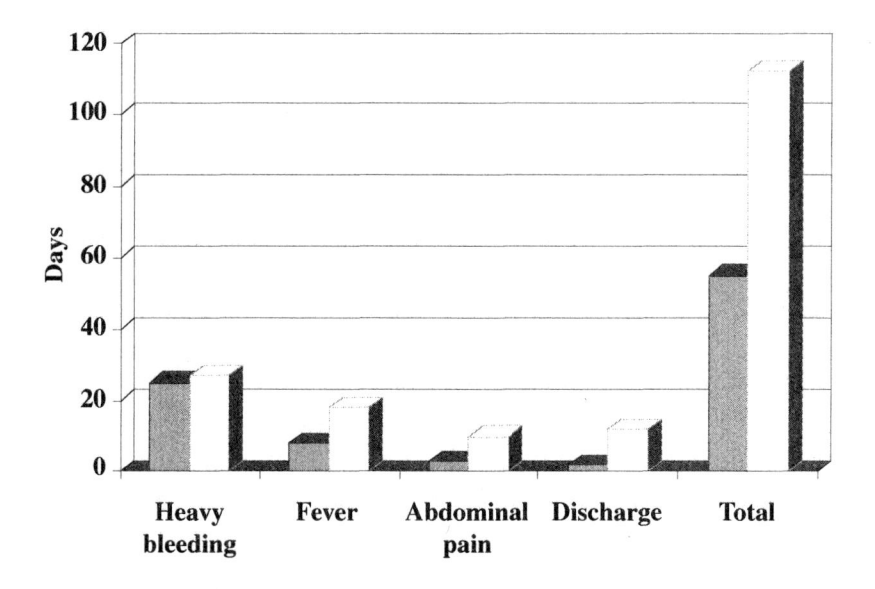

Figure 8.2. *Women Reporting Morbidities by Postpartum Duration*

'batte' (cloth used as a pad). Bleeding in bananthana is viewed positively, as expulsion of bad blood. Excessive bleeding however, judged by having to change cloth pads frequently, big clots and duration exceeding fourteen days, are seen as problems. Even though it is the norm to give 'hot' foods during bananthana, an experienced grandmother stated that, 'After delivery, if bleeding is a lot, food items which cause heat, such as jaggery (molasses) and ganji (wheat porridge), must be avoided. After three months she will get fresh blood and become healthy.'

Traditional bananthana practices are often modified in response to illness and to accommodate health advice. However, sometimes medicines given by doctors are thought to be incompatible with bananthana. For example: Jayanthi had a breast abscess but hesitated to go to a doctor. First she tried herbal medicine. When this did not help and the pain had become unbearable she consulted a doctor, but did not complete the full treatment as it conflicted with traditional care.

Other women who developed breast abscesses tended to avoid medical consultation for fear that their treatment with injections or pills would either dry up their milk supply, or be harmful to the infant.

Another morbidity, fever, is usually thought to be the result of exposure to atmospheric cold, or eating 'cold' foods. Protection from cold, lying down wrapped in a thick blanket, keeping the head covered with a scarf, and putting cotton in the ears is believed to prevent cold from entering the body. Some women say that the abdominal binder and keeping the legs crossed prevents cold from entering through the 'place where the baby was delivered'.

Constipation is common during bananthana, particularly during the early period. The bananthis and the women who conduct the bananthana think it is normal or desirable to defecate only once in a week or once in four or five days. This is partly because it reduces the exposure to the hostile outside environment. Although most women experience pain and discomfort, they do not complain, or view it as a problem.

Three of the respondents reported the incidence of 'beethi shanke' (literally terror and suspicion), the symptoms in these cases being abdominal pain, or strangling sensation, or fever with incoherent speech. This is a locally defined illness to which a new mother and child are thought to be vulnerable. 'Beethi shanke' is thought to be untreatable by modern medical practitioners. This and other related conditions approximate to mental illnesses such as postpartum depression. Local people do not like to say that they or their relatives have beethi shanke, which may indicate that it is stigmatised and under-reported. Even in in-depth interviews, women talk only of other women having this problem, although some admitted, towards the end of the interview, that they had suffered this problem in earlier bananthanas. Certain traditional healers specialise in treating bleeding, breast abscesses and beethi shanke. Some temples are famous for this.

If there is no social support available within the family, care-seeking is restricted. For example, Madevi, who delivered her fourth child, a son, said her banathana was 'worse than that of a dog'. She started to work and use cold water only one month after delivery. When the baby was a few months old, she had pain around the waist, fever and chills. Her husband did not suggest that she should seek care, so she went alone to the hospital, felt giddy and fell over several times. After some time she was able to get someone from the village to accompany her.

In summary, many women who reported morbidity did not seek care, or sought it late. Reasons for this include:

- Traditional restrictions on mobility after delivery
- Beliefs about the aetiology of the symptom, such as bleeding being due to the 'heat' of the post-delivery diet
- Difficulties in accessing care
- Lack of resources
- Lack of support from household members

Nutritional Status in the Postpartum Period

Nutritional restrictions during postpartum in the study area are universally accepted to be beneficial for both mother and child. Unlike some other communities in India, however, there is also a marked decline of activity; women are not usually expected to work, and they are encouraged to keep still. The effect on the woman's nutritional status is likely to be significant. Preliminary analysis of a 'diet and activity' questionnaire administered postpartum to the study women reveals that average daily intake is lower than what is generally recommended. WHO guidelines are:

> Women's intake should be increased to cover the energy cost of lactation, by about 10 percent if the woman is not physically active, but 20 percent or more if she is moderately or very active ... promotion of maternal health has a value of its own. It is important to ensure that women's nutritional status is not undermined by failure to compensate for the demands of lactation.

Although yardstick values for BMI are intended for non-pregnant women, some of the women in the study had low BMI values even when pregnant. Mean weights range from forty-five to fifty kilograms in the predelivery pregnant phase, and these weights, although low, correspond with a number of other studies in south Asia (Kelly et al. 1996). The measurements show a possible risk for poor nutritional status across the postpartum trajectory. Many BMI values were less than 18.4 denoting moderate undernutrition, and some of the women fell within the severe category (BMI < 16.4 [Hutter 1994]). Concern also focuses on the difficulty of regaining weight before the next pregnancy for some of the women. These problems are likely to lead to maternal depletion and the commensurate risks of low birthweight and infant mortality (Hobcraft et al. 1985; Whitworth et al. 2000). Qualitative material gathered with respect to ideal body shapes supported the dietary regimes applied and gave them a rationale based on future suppleness. The following quotes exemplify these views:

> 'A bananthi should become thinner than she was before pregnancy. She should become like the tip of "Mantani" leaf – thin, tender, fresh and supple.'
>
> 'She must become like a sucked out mango stone, like a sliver of wood (chakke).'
>
> 'Her body should become slender and she must become dried and desiccated (sundu hippe).'

The insistence on dryness and being 'sucked out' supports the additional restrictions on liquid intake. Although clearly irksome to breastfeeding women who are often thirsty during their bananthana, there is a high value placed on this aspect of traditional healthy postpartum behaviour. Women feel that this restriction may be difficult, but worthwhile.

Discussion

Beliefs surrounding prescribed behaviour during postpartum (bananthana) are strongly held in this rural South Indian setting, and include restricted mobility, diet, avoidance of ambient and dietary 'cold', and seclusion of mother and child. Medical postpartum care concentrates on family planning advice for tubectomy and immunisation for the child, and the care of the woman herself is popularly seen to be best effected by following traditional care procedures. Indeed 42 percent of the study group had no postpartum contact with medical services if tubectomy contacts are excluded.

Routine postpartum care in this study is determined more by patterns of provision than by care seeking by newly delivered women. Women in village group 2 who had greater odds of receiving routine care were serviced by a dedicated ANM whose house-to-house surveillance is very effective. Even though the actual content of postpartum checks is minimal, the coverage achieved by this ANM is an encouraging example of how the system could be improved. In other areas of the study region, the ANM did not regularly visit, or she was not perceived by women as providing quality care. The patchy nature of rural provision has often been identified as a problem (World Bank 1996), but the perception of quality among local women and their families is also likely to be an important factor where traditional norms do not value medical postpartum check-ups. Furthermore, the reliance of more educated women on problem care rather than routine care points to the need for raising health awareness before postpartum health is popularly recognised as an important part of the childbearing process.

There is generally very little knowledge on the part of local women about the possible types and seriousness of delivery and postpartum problems. This finding concurs with other studies of community awareness of danger signs in developing countries (Adetoro et al. 1991). For example, the terms 'common' and 'normal' are often used interchangeably by women in the study area which means that common conditions are seen as a normal part of pregnancy and accepted without question.

For most of the morbidity reported by the women in the survey there was a care-seeking response, although this was noticeably lacking in the case of heavy bleeding. Postpartum bleeding is a potentially serious morbidity, and which can lead to secondary haemorrhage which can have severe consequences including death (WHO 1998).

Apart from the potentially dangerous conditions of bleeding, fever and signs of hypertension, there is a wide array of postpartum conditions that range from uncomfortable to debilitating and chronic. These morbidities are diverse and women have diverse responses to them. Of concern were the three reported cases of 'beethi shanke'. Most likely, because of the stigma, this is under-reported. Since the completion of the study, a number of reports of women's suicides have been received (8 of the 388 women) and although these were not systematically investigated, we believe that mental health problems may be a cause for concern in postpartum women, as is domestic violence. Clearly a significant proportion of reported morbid events occurred in the post forty-two-day period. Given that this cut off lacks a medical basis, we suggest that *at least* three months be used in assessing maternal morbidity.

A mismatch between biomedical and local views on the correct way to promote postpartum health clearly exists, although there are overlaps whereby traditional practices could be seen by medical practitioners to be beneficial. Increased rest during the postpartum and the intensive attention that is provided by the families of newly delivered women are examples of this. Mobility restrictions, although supportive of energy conservation during bananthana, could be undesirable in terms of the prevention of pulmonary embolism, for which early mobilisation is seen as the most effective prophylactic (WHO 1998). Other traditions, such as reduced food and fluid intake to rid the body of 'bad' fluids and the subsequent emphasis on weight loss could be detrimental. As the data on maternal weight changes show, a significant proportion of women is undernourished, as indicated by BMI values. This raises questions about chronic nutritional depletion, particularly with a first pregnancy that is possibly aggravated by each successive pregnancy.

Steps towards the improvement of postpartum health are clearly constrained by lack of resources, logistical problems and uneven provision. The WHO suggestion of routine postpartum provision at six hours, six days, six weeks and six months after delivery is not currently realistic (WHO 1998). There are, however, aspects of care that are amenable to more immediate improvements. The content of routine care currently emphasises tubectomies and immunisation rather than women's health status. Community acceptance of both problem and routine care would be improved by increased understanding of local customs and beliefs on the part of providers, sensitisation to the need for privacy and a general upgrading of attitudes to women postpartum. In the case of serious or severe morbidity, there is a problem of access to services, transport and referral linkages that needs to be addressed as part of the drive to reduce maternal mortality and serious morbidity. Close attention to the affordability of 'hidden' costs in the public system should also be taken into account when considering uptake of care for serious conditions. The examples of good quality care found in this study and the strong effect of such care on local care-seeking are encouraging and suggest that improvements can be achieved. Previous experience in rural south India has shown that giving information during pregnancy about aspects of postpartum health can affect behaviour, although more culturally unacceptable advice such as postpartum diet guidelines can be more difficult to convey (Nielson et al. 1998). Interventions at the antenatal stage are likely to be effective if cultural understanding is incorporated, and the opportunity to reinforce messages at delivery and at postpartum checks gives great scope for strengthening health promotion. The fairly recent policy framework (World Bank 1996; Government of India 1997) that stresses Reproductive and Child Health (RCH) over the previous emphasis on Family Welfare needs to facilitate the extension of maternal health concerns firmly into the postpartum period. Models of good practice can become an important part of the process of RCH service improvement. With this background, the monitoring of local provision and insistence on basic standards of cover and provision could be a first step to the achievement of good postpartum health.

Acknowledgements

The authors would like to thank the WHO for funding the study, and Dr Iqbal Shah of the WHO and Professor John Cleland of the London School of Hygiene and Tropical Medicine for providing support to the study.

References

Adetoro, O. O., Adeyemi, K. S., Parakoyi, B., Oni, A., Akure, T. and Ogunbode, O. 1991. 'The application of operational research procedures to maternal mortality from puerpural sepsis in a rural community'. *Social Science and Medicine*, 33: 12, 1385–90

Bhatia, J. and Cleland, J. 1995. 'Self-reported symptoms of gynecological morbidity and their treatment'. *Studies in Family Planning*, 26: 4, 203–16

Bhatia, J. C. 1993. 'Levels and causes of maternal mortality in southern India'. *Studies in Family Planning*, 24: 5, 310–18

Defo, B. K. 1997. 'Effects of socioeconomic disadvantage and women's status on women's health in Cameroon'. *Social Science and Medicine*, 44: 7, 1023–42

El-Mouelhy, M., El-Helw M, Younis, N., Khattab, H. and Zurayk, H. 1994. 'Women's understanding of pregnancy-related morbidity in rural Egypt'. *Reproductive Health Matters*, 4: November issue

Fauveau, F., Koenig, M. A., Chakraborty, J. and Chowdhury, A. I. 1988. 'Causes of maternal mortality in rural Bangladesh, 1976–85'. *Bulletin of the World Health Organization*, 66: 5, 643–51

Fortney, J. A. and Smith, J. B. 2000. 'Measuring maternal morbidity', in M. Berer and T. K. Sundari Ravindran, eds. *Safe Motherhood Initiatives: Critical Issues*. Oxford: Blackwell Science

Goodburn, E. A., Gazi, R. and Chowdhury, M. 1995. 'Beliefs and practices regarding delivery and post-partum maternal morbidity in rural Bangladesh'. *Studies in Family Planning*, 26: 1, 22–32

Goodburn, A. and Graham, W. 1996. 'Methodological lessons from a study of postpartum morbidity in rural Bangladesh', in *Innovative Approaches to the Assessmment of Reproductive Health: papers*. Belgium: IUSSP 1996, 257–72

Government of India. 1997. *Reproductive and Child Health Programme: Schemes for Implementation*. New Delhi

Hobcraft, J., McDonald, J. W. and Ritstein, S. 1985. 'Demographic Determinants of Infant and Early Childhood Mortality: A Comparative Analysis.' *Population Studies*, 39: 3, 363–85

Hutter, I. 1994. *Being pregnant in rural south India. Nutrition of women and well-being of children*. Amsterdam. PDOD publication series A. (Doctoral Dissertations)

International Institute for Population Sciences, Mumbai and Institute for Social and Economic Change, Bangalore, NFHS-2 Karnataka Preliminary Report 2000

International Institute for Population Sciences, Mumbai and ORC Macro, USA, NFHS-2 2000

Kelly, A., Kevany, J., de Onis, M. and Shah, P. M. 1996. 'A WHO collaborative study of maternal anthropometry and pregnancy outcomes'. *International Journal of Gynaecology and Obstetrics*, 53, 219–33

Koblinsky, M. A., Campbell, O. and Harlow, D. 1992. 'Mother and more: a broader perspective on women's health', in M. Koblinsky, J. Timyan and J. Gay, eds. *The health of women: A global perspective*. Boulder: Westview Press

Li, X. F., Fortney, J. A., Kotelchuk, M. and Glover, L. H. 1996. 'The postpartum period: The key to maternal mortality'. *International Journal of Gynaecology and Obstetrics*, 54, 1–10

Macro International. 1993. *Indian National Family Health Survey, Karnataka*, International Institute for Population Sciences, Mumbai and Demographic and Health Surveys, Macro International, Washington

Matthews, Z., Ganapathy, S., Ramakrishna, J., Mahendra, S. and Kilaru, A. 2000. *Birth-rights and rituals in rural Karnataka: Care seeking in the intra-partum period*. Social Statistics Working Paper 2000–05. University of Southampton

Murray, C. J. L. and Chen, L. C. 1992. 'Understanding morbidity change'. *Population and Development Review*, 18: 3, 481–503

Nielsen, B. B., Hedegaard, M., Thilsted, S. H., Joseph, A. and Liljestrand, J. 1998. 'Does antenatal care influence post-partum health behaviour? Evidence from a community based cross-sectional study in rural Tamil Nadu, South India.' *British Journal of Obstetrics and Gynaecology*, 105, 697–703

Prual, A., Huguet, D., Garbin, O. and Rabe, G. 1998. 'Severe obstetric morbidity of the third trimester, delivery and early puerperium in Niamey (Niger)'. *African Journal of Reproductive Health*, 2: 1, 10–19

Ronsmans, C. 1996. Studies validating women's reports of reproductive ill health: how useful are they? Paper presented at the IUSSP on 'Innovative approaches to the assessment of reproductive health', Manila, Philippines, Sept. 24–27, 1996

Ronsmans, C., Achadi, E., Cohen, S. and Zazri, A. 1997. 'Women's recall of obstetric complications in South Kalimantan, Indonesia'. *Studies in Family Planning*, 28: 3, 203–14

Rosenfield, A. and Maine, D. 1985. 'Maternal mortality – a neglected tragedy: Where is the M in MCH?' *Lancet*, 2, 83–85

Stewart, M. K., Stanton, C. K., Festin, M. and Jacobson, N. 1996. 'Issues in measuring maternal morbidity: lessons from the Philippines Safe Motherhood Survey Project'. *Studies in Family Planning*, 27: 1, 29–35

Whitworth, A., Smith, P. and Matthews, Z. 2000. 'The health consequences of short birth Intervals, Social Statistics Working Paper'. Department of Social Statistics, University of Southampton, U.K., 20–03

World Bank. 1996. *Improving women's health in India*. Development in Practice series, Washington D.C.

World Health Organisation. 1998. *Postpartum care of the mother and newborn: A practical guide*, WHO-RHT/MSM/98.3, Geneva

CHAPTER 9

'AND NEVER THE TWAIN SHALL MEET':
REPRODUCTIVE HEALTH POLICIES IN THE
ISLAMIC REPUBLIC OF IRAN

Soraya Tremayne

Throughout the world one of the key areas in development planning over the past four decades has been that of population regulation. Most governments have adopted some form of policy to deal with their population size, to either increase or, in the majority of cases, reduce population growth. To achieve this, the state plays an increasingly active role in the reproductive life of its citizens intervening in various ways, directly or indirectly. Some three decades after the first World Population Conference in Bucharest (1974), trial and error, together with a considerable accumulation of new facts emerging from research at an international level, have led to a move away from treating the issue of population from a demographic perspective. The recent paradigm shift from the term reproduction to 'reproductive health' and 'reproductive rights' is the result of this awareness. Human reproduction is no longer treated as an isolated aspect of human life more concerned with numbers and mere economic facts but is instead placed in its broader sociocultural context. There is today universal recognition among policy makers that human reproduction is a complex and intricate process determined by a combination of biological, social, economic and political factors, and that the dynamics of reproduction are unique in each case. At a rhetorical level, planners appear to be aware of the need to understand the underlying cultural factors which determine reproductive behaviour. In practice, however, the gap between rhetoric and reality remains wide. The frequent discordance between the policies devised by health planners and realities at the grassroots level is well documented. In many cases, policies are still more concerned to educate people into their objectives, than to first gain an in-depth awareness of lived experiences on the ground. Furthermore, in many parts of the world, the traditional cultural ideologies are in conflict and compete with the modern rationalities adopted by health policy makers and planners on the basis of international agreements and guidelines. Attempts to accommodate these two worldviews often lead to the adoption of top-down policies, which in turn create further complications and discrepancies in health policies.

In this chapter I will illustrate these points by examining the population policies of the Islamic Republic of Iran (IRI) concerning the reproductive health of adolescents, the compromises which have made their implementation possible, and the result so far. This is not a straight case of modern health policies versus people's traditional values. The situation in Iran is further complicated by the fact that many of the traditional practices and values attached to fertility and reproduction, such as early marriage and high fertility, are encouraged by the jurisprudence (Sharia)[1] of Islam and the Civil Code adopted by the IRI (Draft Country Population Report 1998: 40),[2] Both promote values that are in direct contradiction to those adopted by modern health policies.

The paper is divided into two parts. Part One provides a brief background of the Population and Family Planning Programme of the IRI and its implementation. In the second part I examine the policies adopted in one of the high priority areas of that programme, namely adolescents' reproductive health. The paper will consider the three main parties involved in this process – the conservative authorities who act as the custodians of religious and traditional values, the modernising health-policy makers, and the young people who are the recipients of both ideologies. Since the focus of policies is on the reproductive health and activities of adolescents within the sanctioned rules of the society, the paper will refer only to those areas which concern the 'legitimate' and endorsed reproductive behaviour and activities of young people.

References to Islamic practices in this paper are to Sharia Law, which is universal and followed by all Muslims. It does not examine or compare the differences between the Shia and other Muslim sects.

The Background to the Population Policies

The population policies of the IRI won international acclaim for their success in reducing population growth in a dramatic fashion within a ten-year period, from 1986 to 1996. Iran received the UN Population Award for its achievement in 1998. Through the Population and Family Planning Programme, according to the Statistical Centre of Iran (CSI), the population growth rate, which was 3.8 in 1986, was reduced to 2.5 in 1991 and to 1.5 in 1996.[3] Following this success, the Iranian government gave high priority to reproductive health in its family planning programmes and has steadily increased it budget from $U.S. 2.2 million in 1991 to $U.S. 10.7 million in 1996.[4] The programme's achievement is thus considered to be 'far ahead of the target set in the Second National Five-Year Development Plan'.

Iran's Family Planning Programme was exemplary from its start in that the policies paid special attention to the cultural characteristics of the country and were carefully tailored to social, political and religious rules, including the special requirements of the country's constitution based on Sharia law and religious teachings. By the time the International Conference on Population and Development (ICPD) took place in Cairo in 1994, the Government of Iran had already addressed and integrated the family planning programme into the broader concept of reproductive health and reproductive rights. This was a remarkable achievement, and it clearly demonstrated the awareness of the health-policy makers of the intricacies of human reproduction and its interaction with other social institutions.[5]

The Iranian Delegation to ICPD also played an active part in redefining the Programme of Action adopted at that Conference. However, it did not endorse all the topics included in its agenda. For example, while it agreed with the basic ideals of reproductive health, particularly those concerning the health of women and their right to use contraception for pregnancy prevention, it expressed reservations on some of the topics, inter alia 'the redefinition of the concept of "family" to accommodate non-marital unions', and 'recommendation of the universal exposure of all children to "sex education" programmes at an early age, as contrasted with "an appropriate age"' (Country Population Assessment Report 1998: 3).[6] Also, in spite of the general agreement with the ICPD's Programme of Action, and in spite of the fact that the IRI had in theory adopted a model of reproductive health which covered the whole life span extending from birth to old age, the IRI concentrated its efforts in certain areas of reproductive health only. The Draft Country Population Assessment Report mentions that 'for practical reasons, including lack of resources, all areas and problems of reproductive health are not regarded as of equal concern ... those concerning the health of mothers and children and the provision of contraceptive supplies and services have received much more attention than others'(1998: 4). Other areas of priority included: family planning aimed at enabling couples to make informed decisions about the number of children they wish to have; treatment of disorder resulting from unwanted pregnancies; the promotion of the reproductive health of adolescents, particularly adolescent girls, through education, information and premarital counselling; and the promotion of healthy sexual relations within marital unions.

The Population Policies of the Islamic Republic of Iran

In general, the open-mindedness with which the Iranian government approached the family planning programme was admirable. As Hoodfar (1995: 105) puts it:

> The development of population policy in the Islamic Republic of Iran provides fertile ground for re-examining the widely-held assumptions that Islamic ideology is the antithesis of modernity and incompatible with feminism. By analysing the strategies the Islamic Republic has adopted, in order to build public consensus on the necessity of birth control and family planning, this paper [Hoodfar's] draws attention to the flexibility and adaptability of Islamic ideology to political and economic realities.

After the downfall of the Shah in 1978, and the establishment of the Islamic Republic of Iran, the family planning programme of the previous regime fell into disarray. The new regime did not formulate an explicit policy, and the religious authorities maintained that controlling population growth was a plot by western powers to subjugate oppressed nations and limit the number of Muslims. High population was viewed as the sign of a strong nation and family planning clinics were dismantled. The Iran/Iraq war, which resulted in the heavy loss of approximately one million people, became an important additional reason for the justification of a pro-natalist policy by the government and the religious leaders. The supreme leader of the Islamic Revolution, Ayatollah Khomeini, in his public speeches encouraged people to have large families to replace the 'army of 20 millions'. The new regime also adopted a policy of lowering the legal age of marriage to 9 years for girls and 14 years for boys, according to Islamic teaching.[7] The above were some of the contributory factors for the rise in population growth, which became apparent in the National census of 1986.

The National Census of 1986, which estimated the population at over fifty million, made the authorities realise the enormity of the problem they faced. The population, estimated at 25.7 million in 1966, had almost doubled in the space of two decades. In the light of massive urbanisation, a depressed economy and the demand for jobs and amenities, the high birth rate posed a palpable threat. The Islamic regime had committed itself to the protection of the oppressed and the powerless both ideologically and constitutionally and, unless it could meet some of these undertakings, it would encounter political unrest. It seemed that the government, which was predominantly formed of the clergy, had no choice but to introduce a population control policy, albeit within an Islamic framework. After lengthy negotiations between the religious leaders and planners and policy makers, such a policy was understood and endorsed by a number of the more enlightened religious leaders, who agreed to address the question of overpopulation and its dangers publicly. Ayatollah Khomeini himself discussed the necessity of family planning in Friday sermons, as did the imams of each major city in Iran. By 1988, the Board of Family

Planning, with full support from the highest ranking clergy, had begun work under the control of the Ministry of Health which launched a coordinated campaign aimed at reducing population growth.

In her analysis, Hoodfar relates the success of the campaign to two factors: first, the effort to convince the population of the necessity of family planning, and the launch of a powerful campaign to build consensus; second, the establishment of an effective network to implement the policy of educating the population, especially women, and to provide facilities such as free contraceptives. She distinguishes three broad overlapping themes in the campaign. The first was the international dimension, which addressed the consequences of increased population and argued that high population results in the dependency of the poorer nations on the rich ones. Against this background the clergy (ulama) allowed that Muslims could practice contraception in times of economic hardship, which Iran was now facing due to a revolution, economic sanctions imposed by the outside world, and several years of war with Iraq. The clergy even emphasised the practice of contraception as desirable/prescribed in such circumstances.

The second theme was that of family planning based on reciprocity between the government and people. The campaign pointed out that while no government could legitimately ask individuals not to have any children at all, in exchange families could not have as many children as they wanted since this would put pressure on the government and resources. If Iranians were to build a healthy, educated and able Muslim nation they must find a balance between their individual desires and what the nation could afford. Also, in advocating the methods of contraception, the religious authorities argued that all measures which were temporary and could be reversed (such as pills, intra-uterine devices, tubal legation) were permitted. But, those which led to permanent infertility remained forbidden. Abortion, however remained officially illegal unless the life of the mother was in danger.

The third feature of the campaign was the emphasis on the health of mothers and children. Too many children too closely spaced were considered undesirable for the mother, and the right of children to two years of breastfeeding was highlighted.

In its awareness-raising campaign, the government made funds available for research into Islamic texts to provide proof that family planning had been a concern of Islamic societies in the medieval times and long before the West started thinking about it.[8] The history of Islamic family planning was included in the national curriculum at schools, in adult literacy classes and in local mosques. Special family planning sessions were organised for girls, for male and female workers at the industrial establishments and at urban health clinics. In rural areas, health centres and local mosques were used as a forum for family planning discussions and the media, especially women's magazines, played an important role. Contraceptives such as pills, condoms, IUDs and injections were distributed either free or heavily subsidised. Vasectomy and tubal legations were legalised, although in practice a shortage of trained personnel meant that the latter were not always carried out and the pill became the main method of contraception. The existing health network was strengthened and included health units throughout the country both in rural and urban areas. The health workers, trained especially to work in these units, played a significant part, especially in the rural areas, in the implementation of the Family Planning Programme.

Although the sharp decline in fertility rates was considered to be the result of this well coordinated campaign, and the Islamic Republic took credit for it, there are other views on the factors responsible for the decline. For example, Abbasi-Shavazi (2001) argues that the decline in fertility in Iran started in 1984, when the government was still in the middle of its pro-natal campaign and before the official family planning programme was launched in 1989.[9] In his view, the decline was not the result of the family planning campaign, although without doubt it benefited from the government's capacity to mobilise various organisations and networks and the mass media. Unlike the first pre-Revolutionary campaign launched in 1967, the 1989 campaign was implemented under conditions which were socially, culturally and economically ready to receive it. Moreover, unlike the first programme sponsored by the Shah's regime, the second had the approval and strong commitment of the clergy. The improvement of life conditions, the modernisation of the infrastructure and increased education (especially that of girls) all contributed to the fall in fertility rates. Abbasi-Shavazi argues further that the reduction of infant mortality and the possibility of a better lifestyle for their children encouraged parents to reduce the number of their children. More research is needed to confirm these theories. For example, anecdotal evidence tends to support the view that during the Iran/Iraq war years birth rates increased at all levels, including among some groups of middle income educated people, as well as the more deprived layers of society.[10]

Whatever the reasons behind the remarkable success of the Family Planning Programme, high population growth during the years preceding its implementation in 1986 meant that the 1996 Census population pyramid showed that over 50 percent of the total population of over sixty million, was under 20 years of age, and around 43 percent of this group were under 15 years of age. It also revealed that 10 to 14 year olds form the largest age group in the country with the second largest age group, the 5 to 9 year olds, coming up right behind them (Figure 9.1).[11] The same findings draw attention to the fact that the country should expect a dramatic increase in population in the coming decade when those children, born in the baby boom years of the early 1980s, reach fertility. Some other critical unresolved areas include: the high population growth in rural areas; the unsatisfactory participation of men in reproductive health and family planning programmes; misconceptions about the most effective contraceptive measures such as the pill, IUD, vasectomy and tubal legation, and finally the high percentage of unwanted pregnancies (Figure 9.2). Future population estimates in 1997 were that the total

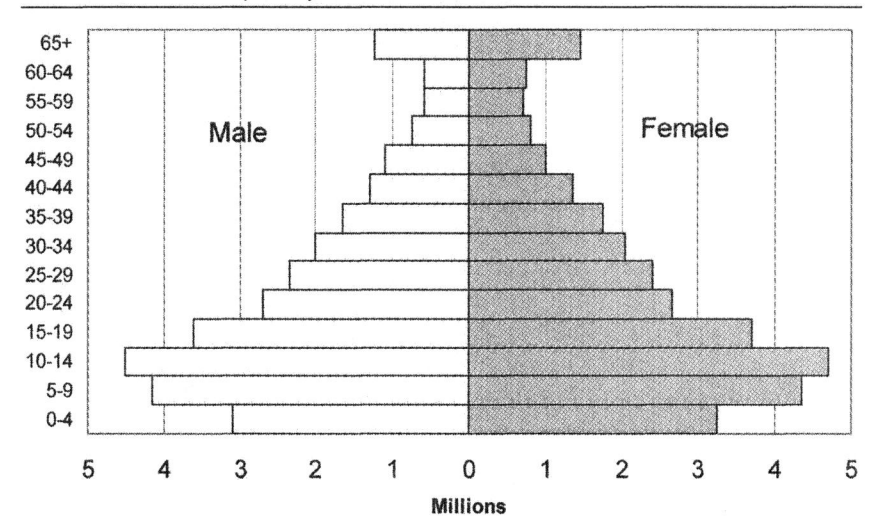

Figure 9.1. *Population Pyramid Based on the 1996 Census*
Source: Iranian Centre for Statistics.

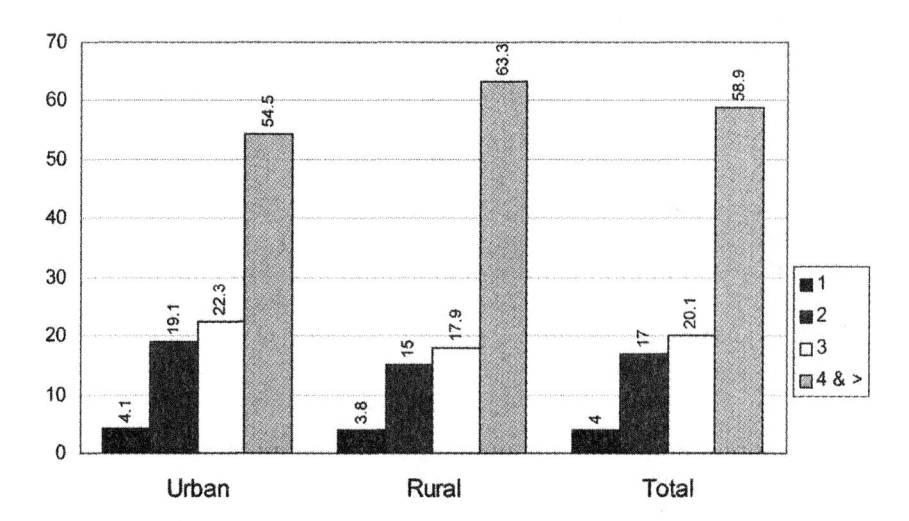

Figure 9.2. *Distribution of Unplanned Pregnancy by Parity in 1996*
Source: Family Health Dept. – Statistic Unit.

population of Iran in 2001 would be 65.7 million, and is expected to exceed 90 million by the year 2021. Although the growth rate of population will decline but only modestly till 2016–2021, during which period it is estimated that the birth rate will average around 1.4 percent per year. The population will therefore stay young for the foreseeable future.

The Modernising Health Policies

In Iran, young people are a dynamic force whose well-being and problems present a formidable challenge to the society as a whole. The authorities are well aware that, unless they can stem population growth, they will have to face major socioeconomic and political problems, and that to control growth they must address the young people's reproductive health. Young people's reproductive health is therefore one of the most important aspects of health planning, and at the ICPD the Islamic Republic undertook the 'Promotion of the reproductive health of adolescents, particularly adolescent girls, through education, information and premarital counselling, and promotion of healthy sexual relations within marital unions.' The concern for 'adolescents' is evident in every official document produced by the government. Various strategies have been developed to address the question of reproductive health among young people, and coordinated efforts are made between the ministries of Health, Education, the Family Planning Association and others[12] to understand and improve their sexual and reproductive health and behaviour. These efforts can be summarised as follows:

Publications: A series of educational text books on reproductive health have been prepared for young people by the Family Planning Association of the IRI. These are written for 10- to 14- and 15- to 19-year-old girls, and 15- to 19-year-old boys, presumably on the assumption that boys under the age of 15 years do not need to be familiarised with reproductive health matters. There are also training manuals and guide books for parents, teachers and other trainers to familiarise them with the biological and psychological facts concerning the reproductive health of adolescents. Most of the literature is based on books prepared by international organisations concerned with population and reproductive health matters, including the International Planned Parenthood Federation (IPPF). These texts are translated and adapted to the requirements of the Islamic Republic by the Family Planning Association of Iran (FPA).[13] The booklets have been issued at schools and other educational institutions. They have also been distributed on a door to door basis in various areas of Tehran. To his surprise, according to the official in charge of the project, the books were better received by people living in the poorer areas, and possibly with less education, than in the wealthier areas where more modern and better educated people live. He interpreted this reaction as 'people from more modest backgrounds respect authority and respond to anybody in an official capacity. They also look up to authorities and are willing to learn from them.' He added that 'people, especially mothers, from the more educated middle classes and from the better off layers of society were reluctant to accept the booklets and said that their children, especially daughters, were already out of control, and they did not want to encourage them to learn more on matters of sex and reproduction.'

Research: A series of research projects have been carried out by the FPA. A substantial qualitative survey (focus group discussion) on the reproductive health needs of 12 to 19 year old adolescents was carried out in 1998 among boys and girls in Tehran. The conclusion in the survey on boys point out that

> ... the knowledge of 15–19 year old boys – the future fathers – in the city of Tehran on matters of puberty, marriage, family planning and venereal disease is very low. In most cases half of the group is totally ignorant, ... Considering that their present attitude and future practice will be a direct consequence of their knowledge and that their knowledge is relatively better than that of the corresponding age/sex group in other cities, it seems there exits a real problem.[14]

A second project was carried out in some of the northern cities of Iran among 10 to 14 year olds, based on focus group discussions. This included parents and teachers as well. The aim was to find out the extent of knowledge among this age group of certain aspects of reproductive health, such as puberty, and also to explore the most suitable channels of communication with young people; who they trust and talk to, and where they get their information from. The survey was ongoing in 2000, and the results not published. The plan was to extend this kind of research to other provinces.

The outcome and recommendations of the existing surveys invariably point to the need to produce more training manuals and books and some counselling for all concerned. The survey samples also seem to be confined to students from schools and other formal teaching organisations. There are no references, for example, to young people at rehabilitation institutions, centres for runaway children, prisons, or counselling committees advising young people on marriage or divorce matters. Furthermore, the surveys are limited to certain aspects of reproductive health, and do not provide a full picture of the reproductive activities of young people.

Health Records: The Educational Division of the Ministry of Health has initiated a new project, to establish a baseline on all aspects of the health, both biological and psychological, of children at school. This is a kind of health log book, which will register children's existing state of health, and monitor their progress throughout their school years. The official view is that this method will allow the ministry to obtain valuable information on all aspects of health, including reproductive health of children. In the summer of 2000, the project was still at an experimental stage and the results had not been analysed.

From the above it can be seen that the main focus of health policies is on educating the young. Implicit in this approach is the assumption that education is the best, or perhaps the only, way to help the young improve their knowledge of reproductive health matters, and that in the long run this will protect them against the hazards arising from their sexual and reproductive behaviour and hopefully also reduce the fertility rates. What is missing from this equation, is the recognition that the target group – 12 to 19 year olds – might already be sexually active within (or outside) wedlock. No official records refer to reproductive activities in the under 15 year age group which take place within marriage and the approved rules of society and with the positive encouragement of the Sharia law. The 1996 Census does not include data on early marriages, pregnancies, or mother and child mortality and morbidity for this age group,[15] although other official documents make several references to their existence. Reference in official surveys is for women over 16 years of age. The more sensitive areas of reproductive health, such as early pregnancies among school girls, HIV/AIDS, sexually transmitted diseases (STD), abortions and other sexual activities, such as homosexuality and prostitution, are deemed sensitive and are, understandably, missing.[16] This is not unique to Iran; the majority of states prefer not to release information on these issues.

Legitimate but Absent Reproductive Practices

We mentioned earlier that the ungrudging support of religious leaders gave legitimacy to various methods of family planning, and the condition of this consent was that the programme should act within an Islamic framework. Here we examine the traditional and religious understanding of reproductive health and the boundaries the health policies are permitted in their planning in this area. When the policies refer to the reproductive health of adolescents, apart from designing schemes for educating them, they invariably raise the issue of early marriages as the only existing form of reproductive activity for this age group as if no other activity of this kind exists or is recognised and accepted. For example, according to the Draft Country Population Assessment Report (1998: 40), 'as part of the genuine commitment to improving the reproductive health of young people, especially girls, the Civil Code in recent years has raised the age of marriage to 15 and 18 for girls and boys respectively'. In practice however, there is to date no definite lower limit for the age of marriage other than that prescribed by Islam, and the efforts of Islamist female members of Parliament to raise the age of marriage have met with strong resistance, mainly from the conservative male members (debates in the Iranian Parliament, October 2000). The existence of early marriages is also openly acknowledged:

> Because of the emphasis of Islam on early marriage and the large number of adolescent brides and grooms, the reproductive health needs of young married couples requires serious attention by the family and health system. Over the past few years a number of steps have been taken to address these particular needs of youngsters intending to get married. (The Draft Country Population Report 1998: 44)

To achieve this, premarital counselling committees have been formed by the government to advise young couples on reproductive health matters, family planning and child spacing. Although the original aim was to try and delay early marriages through counselling, in practice, and ironically, some of these committees are used as advisory committees by people who intend to marry their young children (both boys and girls). Counsellors in some of the more conservative areas of the country find themselves faced with having to advise 10 to 12 year olds brought to them for marriage counselling.[17] This is further illustrated by the activities of various traditional and/or religious organisations aimed at encouraging traditional practices such as early marriage. For example, the charity organisation known as the 'Imam's Relief Committee', which periodically organises mass marriage ceremonies for the economically deprived youth under its care has established a premarital family planning education and counselling service for these young couples (40).

As well as the official sources which acknowledge the existence of early marriages, there is abundant evidence of them in the daily papers. One such report, by no means rare, recounts a case referred to the police of a 10-year-old girl and her mother beaten up by the girl's husband for refusing to do the housework because the girl wanted to play with her dolls.[18] A further confirmation of the existence of early marriages, which is also given as proof of the commitment to the education of women, is that the National Education Policy mandates schooling for married pregnant girls.

Official records exist for marriages of 16 year olds and above; but no specific data exists for the age group below. Nevertheless, the indications are that not only do early marriages exist, they may be even on the increase, as is the case in the rest of the region (UNICEF March 2001).[19] An UNFPA report,[20] for example, draws attention to the fact that statistics showing that the age of marriage is on the rise may disguise the continued practice of early marriage in certain areas or among certain population groups. According to this report, early marriage is apparently increasing among populations under severe stress – in conflict situations, confronted by the HIV/AIDS epidemics, or facing extreme poverty. In Iran the age of marriage has gone up considerably in many areas, but the extent of the practice of early marriage varies in different provinces and among different ethnic groups. In spite of the lack of data on early marriages, it would perhaps be a safe guess that in those areas where they were traditionally practised extensively – in the less developed and more conservative parts, and where there is perhaps more continuity than change as far as family values and fertility and reproductive practices are concerned – they still exist to a considerable extent.[21]

Ethnographic research carried out before the 1979 Revolution includes a wealth of information on early marriages. To mention one such study, research among the Bahmei tribes of Kohkiloye (a Province in central south west Iran) revealed that from a sample of 372 women interviewed, 75 percent had married before they were 15 years old (82/83).[22] Kohkiloye was thought of as one of the most remote areas of Iran in 1970s, and the existing socioeconomic indicators show that it remains one of the least developed and most conservative and traditional parts of the country. The consistency of the socioeconomic indicators between past and present may be one indication that reproductive health practices also remain consistent. This view is further substantiated by the Draft Country Population Report (1998: 38) which mentions the considerable gaps that exist in the rates of maternal mortality and morbidity between different provinces. The Report gives detail of prevalence and cites as an example the percentage of births unattended by trained personal in some rural areas such as Sistan and Baluchestan (80 percent), Kohkiloyeh and Boyerahmadi (61 percent), and Hormozgan (56 percent). A final insight into the prevalence of traditional family structures comes from research on female suicide (usually by self-immolation) in Kohkiloye, where the rate is among the highest in the country. The persistence of strong patriarchal values, and their clash with the changes (albeit limited) brought about by development are cited by some as the causes of these suicides.[23] Similar observations could be extended to some other provinces such as Ilam and Kurdistan.

In spite of the absence of reliable data, all the indications point to the practice of early marriages, and an in-depth investigation of the extent as well as the context in which they happen is crucial to throw new light and open new avenues for health policies.

The Adolescents

The agenda of health-policy makers is to modernise societies with the general wellbeing and 'improvement' of life conditions in mind. Opposition to modernisation and development usually comes from the conservative forces in society. These too have the 'wellbeing' of people in mind and are anxious to achieve it by trying to preserve the continuity of traditional cultural practices. In the case of population and health planning in Iran, the clash between tradition and modernity does not begin or end with planners versus people, on a straight line with two forces pulling in two different directions. There is a third force, namely the massive group of adolescents, the recipients of both ideologies, which is changing the shape of the confrontation into a triangle.

The young form a formidable and dynamic force in Iran, with 50 percent under 20 years of age, 43 percent of these under 15 years of age, and 27 percent (nearly sixteen million) in the 10- to 19-year-old age range. Overlooking the importance of in-depth research on the social and cultural factors which determine sexual and reproductive practices can but contribute towards their perpetuation. Devising policies on the assumption that adolescents are passive consumers of policies, and hoping that through coercion (by the conservative forces) or education (by the health planners) only, the aims of health policies will be achieved goes beyond the bounds of optimism and any vision these ideologies may have for resolving the question of high fertility growth. Their project, in other words, becomes utopian.

Notwithstanding socioeconomic and cultural differences, and ethnic and regional variations, this massive group of young people has one thing in common: its exposure to the outside world through education, urbanisation, the media and the general process of globalisation, which have created certain common values and aspirations among them. Young people play an active role in both reconstituting the traditions they inherit and in negotiating the modern values which are imposed on them. They react in a variety of ways to both ideologies, if these do not fit in with their own social realities and ideals. Even though they may not always have the power to fully control their lives, they use whatever means are available to them to find ways of reshaping them. Their reactions vary and can generate a range of responses from inventiveness and imagination, to resourcefulness and adaptation. There is abundant evidence that in Iran the social world of young people does not necessarily match that of either authorities, and this is reflected in a variety of ways. It is, however, beyond the scope of this paper to elaborate on this fascinating issue. In what follows I shall concentrate on the fundamental cause of the polarity of reproductive ideologies which has resulted in a widespread malaise among all concerned.

Age Categories and Gender Roles

The persistence of traditional reproductive practices, and the obvious reluctance of policy makers to admit that there are 'children' who are married and have children, and/or are, 'legitimately' or 'illegitimately', sexually active, is not just a matter of the dichotomy arising from the struggle between the traditional values and modern practices. Society has its own understanding of the concept of age, and of gender roles within this. Age categories and gender relations are closely intertwined and both reflect and determine the ideology a community holds of its identity and reproduction.

The understanding and interpretation of age categories vary greatly across cultures and remain a universal dilemma for policy makers. The rapid changes in transitional societies have created an inevitable clash between the modern and the traditional worldviews. Factors referred to earlier such as urbanisation, globalisation, the influence of the media and the increase in consumerism, combined with better access to education, and improvement in some aspects of health, have created similar trends worldwide. The apparent similarity of these trends has been taken as a cue by policy makers to search for universal solutions. In an effort to respond to the needs of different groups, universal age categories such as 'children', 'young people', 'adolescents', 'youth' and 'young adults', have been identified, without differentiating the specific social and cultural contexts which determine these categories. The discrepancies which exist between the international definitions of children, young people, teenagers, and other age categories on the one hand, and local practices and values as far as life trajectories are concerned on the other hand, are creating constraints at several levels. What definitions of this kind do not take into account is the extent to which such terms are social and cultural constructs grounded in their unique historical and economic settings (Harcourt (1997: 11).[24] For example, while according to the Convention of the Rights of the Child (CRC) anyone under the age of eighteen is a child, in many developing countries married 12-year-old girls are considered adults, while a 17-year-old boy at school is considered a child.[25] Likewise, the concept of adolescents needs clarification in its specific context. Indeed it is a relatively new one in most cultures. Even in the West it has a history of a hundred years or so and has been introduced to non-Western cultures exposed to development through the process of modernity (Harcourt (1997: 12).

The 1989 Convention on the Rights of the Child (CRC) defines a child as anybody under the age of eighteen. Although the IRI is a signatory to this Convention, it has reserved the right not to apply some of the clauses if they contradict Islamic standards and internal laws and practices of the country. For example, the Iranian delegation to the CRC maintained that the Sharia law provides clear instructions for the responsibilities and conduct of people, and in some instances these do not accord with international agreements. A review of the Iranian laws concerning children reveals that there is no clear definition of the end of childhood; a child can be considered a child in one context and an adult in another. For example, a 10-year-old girl is considered to have reached puberty (according to Islam) and therefore to be eligible for marriage and commercial transaction. However, the same girl cannot be employed according to the labour law which stipulates a minimum age of 15, or obtain a passport or driving licence. Although a 9-year-old girl can get married and have children, she is not eligible to vote (the age of voting is currently set at 15), in spite of the fact that she has the status of a full adult. In the area of criminal law a 10-year-old girl and a 15-year-old boy are liable to sentences similar to a full adult. In theory they can be sentenced to death, although in practice such a sentence has not happened. In a case of theft, a 9-year-old girl and 15-year-old boy can be sentenced in the same way as a 40-year-old man. A recent report from UNICEF mentions that the minimum age for sentencing children has now been raised to fifteen.[26]

Adjusting the Islamic worldview of age to recent changes in the position and status of children is a difficult process and it will take some time for these filter down into the traditional and conservative values and alter them. The obvious conflict between the two sets of values is summarised in the findings of the survey on Adolescents' Reproductive Health Needs Assessment, referred to earlier, which reveals the resentment of young people about their confused status in society. The young lamented that 'generally there is a lack of respect for young people from parents and teachers in the society'; that 'adults don't pay enough attention to our ideas'; that 'there is a lack freedom, especially for girls, and restrictions in making contact with the opposite sex'.

Gender

Concepts of gender and gender roles are closely linked to age, universally. Islamic scholars have debated the issues of gender and Islam exhaustively, and I shall not elaborate in detail on them. In summary, in Islam, women are viewed as responsible for the protection of family and the preservation of its rules, 'in order to protect the society from the evil of the breaking up of the sexual barriers'. The traditional view of marriage is directly linked to the formation of family; the desirable age of marriage and the way in which a spouse is selected depend on a society's view of the family, its role, structure, pattern of life, and the individual and collective responsibilities of its members. The idea and function of 'family' varies across the world and is in a state of constant evolution. Traditionally, marriage and the formation of marital unions in the patriarchal Muslim societies were viewed as a transaction, with female sexuality as a commodity traded by men. In an essay on female sexuality, Moghadam (1994) addresses the issue of its ownership and argues that the legal commoditisation of female sexuality should be incorporated into the analysis of gender economics in Muslim societies.[27] Moghadam defines the concept of commodity as follows: 'an object that is voluntarily bought and sold at an agreed price is a commodity. Once the sale transaction is completed, the original owner loses ownership and the buyer becomes owner of the commodity. With a few modifications these definitions could be used for the discussion of female sexuality in Islam' (1983: 83). Haeri (1989) also argues that in Islam permanent marriage is a sale and temporary marriage (mainly practised among the Shi'ite Muslims) is a lease of female sexuality.[28] In her analysis of marriage, Mir Hosseini (1993: 1) in addition considers the two aspects of Islamic law: the 'temporal' and the 'religious'. She argues that 'the boundaries of 'sacred' and 'temporal' are blurred in the case of family law, for which divine revelations are abundant. 'Islamic family law is permeated with religious ideas and ethical values; it holds within itself a distinct model of family and gender relations. This model is claimed to be divinely ordained and, thus, immutable.' Mernissi (1975) too argues that in Islam the male-female relationship is assumed to be highly sexual, and that female sexuality needs to be kept under control.[29] While there are very few areas of prohibition on the sexual activities of men, whether before or after marriage, sexual prohibitions abound for women and concern them in the main. Development has brought about fundamental changes in the lives of women. Female education, general improvement in women's health, and above all the increasing recognition of the value of women's economic contribution and their participation in the labour market, have led to a certain freedom and power to control their own lives. These changes have restricted some of the control that men have had on women, and have left them in a state of bewilderment. The consequences of such perplexity are reflected in a drastic increase in the rate of divorce, violence against women, and general social disorder, resulting in the shattering of family structure in Iran.[30] An increasing number of girls, under pressure and suffering violence from the male members of their family, choose to run away from home, and many (both boys and girls) end up in prostitution or get caught up in the trafficking of children. As Boyden (2001) puts it 'along with concerns about violence, crime, and subversive youth cultures, sexuality and reproduction are the spheres of youth activity and attitudes that provoke the greatest moral panic and receive the greatest attention from policy-makers.'[31]

Policies at Odds with Themselves

Considering the genuine concern for and the amount of effort and planning put into the reproductive health of adolescents, it is ironical that it is the realities of the life of this very age group, and its reproductive experiences, which are left out of official data altogether. The unease with which the reproductive health policies approach the issue mirrors the tension between conservative and reformist forces in the society. The risks of early marriages and negative effects of early pregnancies on the physical and mental health and schooling of adolescents are wellknown to health-policy makers. The dilemma created by having to define their policies within an Islamic framework, and to accommodate what they themselves judge to be 'right', has therefore been solved through compromises, made at the cost of any acknowledgement of the sexual and reproductive health activities of the largest age group in the country; by, in other words, turning a blind eye to the actual sexual and reproductive life of young adolescents. On the one hand, any form of sexual and reproductive activity outside wedlock for young people is considered 'pathological' and a real threat to the stability of society by the conservative forces. Premarital sex and reproduction by adolescents, especially girls, is socially prohibited and subject to severe punishment (Draft Country Population Report 1998: 40). On the other hand, the truth about the sexual and reproductive life of young adolescents, even their legitimate activities, seems to be too embarrassing for the modernising policy makers to include in their data. The traditional practice of becoming sexually active at an early age under the weight of tradition and religion is deemed 'pathological' by modern standards, as is promiscuous sex because of the health dangers it represents. Therefore, the compromise adopted by planners is that this age group (under 15 years) has not yet entered sexually active life, and all it needs is education to familiarise it with its reproductive functions. Consequently, the issue of adolescent sexuality and problems associated with it are somehow ignored either by society and/or the health system. Health planners use innuendo to present the facts without being seen to put their seal of approval on them. The diffident tone of some of the health policy documents testifies to this search for acceptable ways forward. Young people, who are at the receiving end of conflicting messages, remain confused as to their identity and status in the society. Their behaviour is considered 'pathological' by both sides. Disagreements and contradictions between the two sets of values abound, and the sensitivity of the subject remains one of the major obstacles with regards to adolescent reproductive health.

There are two further obstacles in this process. First, while some information and insight into the sexual and reproductive life of young people does exist, it is incomplete. Little effort has been made to analyse what material is available or to provide quantifiable material for the policy makers to include in their data. It is often inherent in the nature of policy making at a macro level that, unless facts can be quantified, they are of little consequence and not worth planning for. Information at a micro level is rarely treated with respect or considered relevant. The second factor is indirect resistance on the part of many of those responsible for planning or the implementation of plans, especially those who belong to the more conservative layers of society, who themselves remain unconvinced by modern policies. Many of these people, at heart, support the old values of family and fertility.

Conclusion

The preceding discussions suggest that in designing and implementing development projects, discord arises between the forces of conservatism trying to preserve the national identity, rooted in tradition, and policy makers who seek to 'better' life conditions through modernisation. In the case of the population and family planning programme of the Islamic Republic of Iran, the usual pattern of the state versus people is further complicated by the presence of a third force, young people whose reproductive health is identified as one the most important areas of health planning, and who are caught in the conflict between the state and its planners. The underlying factors responsible for the clash between the state and its policy makers are, above all, due to the fundamental and ideological differences between the two worldviews of society, of family structures and marriage, of age, and of gender roles within this, between conservative and modernists. The question arises as to why the same authorities, who agreed to the implementation of an enlightened and successful family planning programme during its first phase, have reverted to an Islamic framework during the second phase, and are now creating restriction for policy makers? The case of the population and family planning of the IRI is a clear example of the interaction between politics and reproduction.

While no attempt has been made to develop a political framework for the analysis of the reproductive health policies of the Islamic Republic in this paper, it hints at the clash between the agenda of the religious political elite and that of the modernising bureaucrats. During the years of rapid population growth, the religious leaders, as political leaders, agreed to a population control programme which they justified in the name of Islam to build up a strong, well-educated and healthy nation. To ignore population growth would in reality have meant a serious political threat to the regime and its ideology. The first phase of these population policies went ahead relatively smoothly. But, at a later stage when it became necessary to gain a real understanding of reproductive practices as lived at the grassroots, inconsistencies and conflicts between the two ideologies began to emerge. The conservative and religious leaders continued with their active support for traditional family structure and sexual and reproductive practices. In the meantime, modernity, inevitably, has disrupted these systems, and alienated young people from the traditional systems of support. With the change in life conditions, young people no longer have the opportunity to learn from their parents and grandparents, and the new channels of information such as school, media and the peer group can only provide partial information.

The authorities' (both conservative and modern) method of treating young people as children and ignorant and in need of teaching and guardianship, while they are expected to play several roles as both adults and children at the same time, have created a confusing situation for all concerned. Evidence shows that young people, of any age, play an active part in shaping their own lives, and that their behaviour represents a wide range of experiences, which go beyond the 'pathological' and demonstrate a profound desire to assert their existence.

Several themes emerge from the discussion: (i) attempts to devise policies for young people's sexual and reproductive life without a profound understanding of the social and cultural context in which they live are a barren exercise; (ii) a dimension often overlooked in the implementation of policies is that the attempt to teach the young the biological facts of reproduction in the hope that they will reduce their fertility does not necessarily lead to a reduction in the number children they wish to have. Decisions on sexual and reproductive activities are not easily influenced by outside factors, unless these correspond to people's own social realities and ideals; (iii) that the encounter of top-down policies with local realities tends to provoke reactions which may not be direct and active, but passive and indirect, often frustrating policies without being seen to be confrontational. In the case of Iran, active confrontation of young people against either old or modern values is on the increase in most spheres of life.

The example of Iran is by no means unique, and will be familiar to many researchers, policy makers, implementers and to the young themselves throughout the world. The analysis of the Iranian case is used to unravel the particular model of interaction when policies meet local realities. In another context, the components may be different, but will the result differ? The question remains: will the encounter of policy with local reality ever produce the 'desired' effect? Or, will it continue to provoke reactions and create new and unforeseen situations, positive or negative? Perhaps the answer can be found in the following passage by David Parkin (2001):

> the management of reproductive life is not just a matter of formal systems of authority, control and knowledge transmission, nor of following

policy-makers directives, nor even integrating these with indigenous ones. It also extends to recognising the metaphysical and cosmological understandings and practices that everywhere accompany rules, plans and policies.[32]

Notes

1. Although there is a distinct difference between the *Sharia* (Revealed Law) and *Fiqh* (jurisprudence), the lines between the two are blurred, as Mir Hosseini explains (1993: 5) 'Muslims do not normally distinguish between the two, especially at the popular level'. In Mir Hosseini's words 'a distinction between the two is central to the concept of law in Islam. *Sharia* is law in the sense that it contains a divine blueprint for mankind, for both this world and the other. It is the Will of God for humanity. *Fiqh*, which contains the letter of law, is "the whole process of intellectual activity which ascertains and discovers the terms of divine will and transforms them into a system of legally enforceable rights and duties".' (Coulson, 1969: 2, cited in Mir Hosseini (5). For more details see Mir Hosseini, Z. (1993), *Marriage on Trial: A Study of Islamic Family Law*. London: I.B. Tauris.
2. Draft Country Population Assessment Report of the Islamic Republic of Iran (1998). Drafted by the Plan organisation of Iran. Unpublished and with limited circulation
3. Different sources give different figures for the population growth rate, but in this chapter I shall use the figures given by the Statistical Centre of Iran, as used in the Draft Country Population Assessment Report 1998.
4. The Country Report on Population, Reproductive Health and Family Planning Programme of the Islamic Republic of Iran, 1998. Prepared by Family Health Department. Under Secretary for Public Health. Ministry of Health and Medical Education. Islamic Republic of Iran.
5. Hoodfar, H. (1995) has written extensively on this topic. See, for example, 'Population Policy and Gender Equity in Post-Revolutionary Iran', in C. Makhlouf, ed. *Family, Gender and Population in the Middle East: Policies in Context*. Cairo: American University of Cairo Press, 1995. See also Makhlouf, C. (1994) 'Reproductive Choice in Islam: Gender and State in Iran and Tunisia', *Studies in Family Planning*, vol. 25/1, January/February 1994: 41–51.
6. Draft Country Population Assessment Report of the Islamic Republic of Iran (IRI), 1998.
7. See also Abbasi-Shavazi, M. J. (2001), 'La fecondité en Iran: l'autre revolution'. *Population et Société*, no 373, November 2001.
8. See also Makhlouf, C.'s for an explanation of the history of family planning in Islam in 'Women, Islam and Population: Is the Triangle Fateful?' Working Series Paper No. 6, Harvard: Harvard School of Public Health, *Harvard Centre for Population and Development Studies*. (1991: 42–43). As Makhlouf explains: 'a clear consensus exists among all schools of Islamic law that family planning is permissible. This is based both on the absence of any prohibition against birth control in the Koran and on general statements in the Koran that God does not want to burden man but wishes to improve his life. More specifically, statements in the Hadith indicate that withdrawal (*azl*) was practiced in the Mohammad's time and that he did not discourage his followers from the practice'. As Makhlouf points out, the Sunni and Shia position on birth

control are in substance the same. They derive from the work of Al-Ghazali, the most celebrated theologian of Islam, who establishes five reasons for which birth control may be allowed.
9. This view is also supported by Draft Country Population Assessment Report: 21
10. The assumption cannot be made that the rise in population was happening among the poorer layers of the society only, nor that the prospect of a better future had made people reduce the number of their children. Interviews with several middle class and educated families in Tehran revealed that they had increased the number of their children. They explained this by saying that 'We have just had a revolution and there is the war with Iraq. As a result we are all out of jobs. All this will come to an end sooner or later, and we shall eventually return to work. It is a good idea to have several children now, so that by the time everything returns to normal our children will be grown up, and can help us'. Personal interviews in Tehran 1984/5.
11. Country Report on Population, Reproductive Health and Family Programme in the Islamic Republic of Iran, 1998. Ministry of Health and Medical Education, Islamic Republic of Iran.
12. Several of the UN agencies especially the UNFPA, WHO, UNICEF, and IPPF have played an active role and closely co-operated with the Iranian authorities in their population and family planning programme.
13. The Family Planning Association of the IRI, which is the Iranian branch of the IPPF, has been the main body writing and translating informative and teaching materials, and also carrying out research on young people.
14. Reproductive Health Needs Assessment of Adolescent Boys, 1998. Family Planning Association of the Islamic Republic of Iran: 23.
15. Draft Country Population Assessment Report 1998. Also during an interview with officials at the Ministry of Health, it was stressed that marriages or pregnancies among the under 15 age group do not exist. Interviews in the summer of 2000 with marriage counsellors in some of the provinces, on the other hand, confirmed that these are practiced and are not rare incidents.
16. For reference to the sensitivity of the issue elsewhere, see, for example, Mensch. B., W. H. Clark, C. Lloyd. and A. S. Erulkar, 'Premarital Sex, Schoolgirl Pregnancy, and School Quality in Rural Kenya'. *Studies in Family Planning*, vol. 32 no. 4, December 2001.
17. Personal interviews with counsellors August 2000.
18. *Keyhan*, London, October 2001
19. Early Marriage: Child Spouses, UNICEF, Innocenti Research Centre, Florence No. 7 March 2001.
20. Nafis Sadik, Executive Director, UNFPA. 'Working Towards Gender Equality in Marriage' in Innocenti Digest, UNICEF, No. 6, Domestic Violence.
21. A recent report (April 2002) issued by the Chief of Police for Greater Tehran mentions that the number of runaway children is increasing drastically, and that in the last four months alone 6,156 children have run away from home.
22. Restrepo-Afshar Naderi, E. (1975) 'Le Mariage dans la Tribu Bahmei (Kohgiluye-Iran)' unpublished thesis (1975), Université de Paris VII U.E.R. Anthropologie, Ethnologie, Science des Religions. In addition, there exists a wealth of ethnographic research carried out before the 1979 Revolution, by Iranian and foreign researchers, which *inter alia* include ample references to the practice of early marriage. For example see the series of monographs published by the Institute for Social Research of the Faculty of Social

Sciences, University of Tehran, pre the 1979 Islamic Revolution. These studies were led by Keshavarz, H. for the rural areas and by Naderi Afshar, N. for the tribal and pastoral nomads.
23. The main source of information is research carried out by the sociologist Jaleh Shadi Talab. Various sources, too numerous to cite here, confirm these findings. One might also point to the many articles published in Iranian journals and magazines that alert the authorities to these problems.
24. Harcourt, W. A. (1997: 11) 'An Analysis of Reproductive Health: Myths, Resistance and New Knowledge'. *Power, Reproduction and Gender: The International Transfer of Knowledge*. London: Zed Books, 8–34.
25. For more details on this subject see 'Introduction' in S. Tremayne, ed. *Managing Reproductive Life: Cross-Cultural Themes in Fertility and Sexuality*. Oxford: Berghahn Books, 2001, 17–29.
26. *The Situation of Children and the Young in Iran*, report prepared by the Iranian Office of UNICEF, 1998.
27. For more information on this topic see Moghadam, F. (1994) 'Commoditisation of Sexuality and Female Labour Participation in Islam: Implications for Iran', 1960–90, in M. Afkhami and E. Friedl, eds. *In The Eye of the Storm*, London: I.B. Tauris.
28. Haeri S. (1989) *The Law of Desire: Temporary Marriage in Iran*, London: I.B Tauris.
29. Mernissi. F. (1975) *Beyond the Veil: Male-Female Dynamics in Modern Muslim Society*. Bloomington: Indiana University Press. See also Afkhami, M. (1994) in M. Afkhami and E. Friedl, eds. *In The Eye of the Storm*. London: I.B. Tauris.
30. Iran is considered to have one of the highest rates of divorce in the world after the United States. One in every 10 marriages ends in divorce.
31. Boyden. J. (2001) 'Some Reflections on Scientific Conceptualisations of Childhood and Youth', in S. Tremayne, ed. *Managing Reproductive Life: Cross-Cultural Themes in Fertility and Sexuality*. Oxford: Berghahn Books, 2001: 175–93
32. Parkin, D. (2001) 'Foreword', in S. Tremayne, ed. *Managing Reproductive Life: Cross-Cultural Themes in Fertility and Sexuality*. Oxford: Berghahn Books.

References

Abbasi-Shavazi, M. J. 2001. 'La fecondité en Iran: l'autre revolution', *Population et Société*, 373
Afkhami, M. 1994. in M. Afkhami and E. Friedl, eds. *In The Eye of the Storm*. London: I.B. Tauris
Boyden, J. 2001. 'Some Reflections on Scientific Conceptualisations of Childhood and Youth', in S. Tremayne, ed. *Managing Reproductive Life: Cross-Cultural Themes in Fertility and Sexuality*. Oxford: Berghahn Books
Draft Country Population Assessment Report of the Islamic Republic of Iran. 1998. Plan Organisation of Iran. Unpublished and with limited circulation
The Country Report on Population, Reproductive Health and Family Planning Programme of the Islamic Republic of Iran. 1998. Prepared by Family Health Department. Under Secretary for Public Health. Ministry of Health and Medical Education. Islamic Republic of Iran
Early Marriage: Child Spouses, UNICEF, Innocenti Research Centre, Florence No. 7 March 2001

Haeri, S. 1989. *The Law of Desire: Temporary Marriage in Iran*. London: I.B. Tauris.

Harcourt, W. A. 1997. 'An Analysis of Reproductive Health: Myths, Resistance and New Knowledge' in *Power, Reproduction and Gender: The International Transfer of Knowledge*. London: Zed Books, 8–34

Hoodfar, H. 1995. 'Population Policy and Gender Equity in Post-Revolutionary Iran', in C. Makhlouf, ed. *Family, Gender and Population in the Middle East: Policies in Context*. Cairo: American University of Cairo Press

Keyhan, London, October 2001

Makhlouf, C. 1991. 'Women, Islam and Population: Is the Triangle Fateful?' Working Series Paper No. 6, Harvard: Harvard School of Public Health, Harvard Centre for Population and Development Studies, 42–43

Mensch, B., Clark, W. H., Lloyd, C. and Erulkar, A. S. 2001. 'Premarital Sex, Schoolgirl Pregnancy, and School Quality in Rural Kenya'. *Studies in Family Planning*, 32: 4

Mernissi, F. 1975. *Beyond the Veil: Male-Female Dynamics in Modern Muslim Society*. Bloomington: Indiana University Press

Mir Hosseini, Z. 1993. *Marriage on Trial: A Study of Islamic Family Law*. London: I.B. Tauris

Moghadam, F. 1994. 'Commoditisation of Sexuality and Female Labour Participation in Islam: Implications for Iran, 1960–90', in M. Afkhami and E. Friedl, eds. *In The Eye of the Storm*. London: I.B. Tauris

Nafis Sadik, Executive Director, UNFPA. 'Working Towards Gender Equality in Marriage', in *Innocenti Digest*. UNICEF, No. 6, Domestic Violence

Parkin. D. 2001. 'Foreword', in S. Tremayne, ed. *Managing Reproductive Life: Cross-Cultural Themes in Fertility and Sexuality*. Oxford: Berghahn Books

Reproductive Health Needs Assessment of Adolescent Boys. 1998. Family Planning Association of the Islamic Republic of Iran: 23

Restrepo-Afshar Naderi, E. 1975. *Le Mariage dans la Tribu Bahmei (Kohgiluye-Iran)*. Unpublished thesis. Université de Paris VII U.E.R. Anthropologie, Ethnologie, Science des Religions

Tremayne, S., ed. 2001. *Managing Reproductive Life: Cross-Cultural Themes in Fertility and Sexuality*. Oxford: Berghahn Books, 17–29

CHAPTER 10

WOMEN IN FERTILITY STUDIES AND *IN SITU*

Tulsi Patel

This paper discusses one of the key concepts in demography, especially fertility studies. The focus is on the concept of 'status' of women. 'Status' has been used in sociology and social anthropology from where it seems to have found its way in demography. The paper explores the facility and yet the limitation of the application of 'status'. Though the concept of status is in some disuse even in sociology, its usage towards reproduction (of difference and distinction) is seen in recent times (Bourdieu 1986). The first section traces the centrality of women in demographic studies. The next section looks at the application of status by some of the demographic studies. The third attempts a critique of these studies on empirical considerations. The preference for the terms 'autonomy' and 'empowerment', both implying isolated and individual agency, is also alluded to. The fourth section tries to explore the conceptual and methodological assumptions and disciplinary demands that pose some analytical challenges to this genre of research. The relevance of the concept and the lessons learnt from the research are in the conclusion. The data from my study in rural Rajasthan, a state in northwest India, are provided intermittently.

Status is the relative standing or position in society seen either in the Marxist or Weberian tradition in sociology. For Marx, economic characteristics based on ownership of private property form the basis of social stratification and the ultimate divisions in society, while Weber assigned primacy to power, economic and cultural differences. The relative standing involves conferring a bundle of socially defined attributes, rights and obligations upon a person occupying a position and having a social role in society. Status connected with a social role is, rights and expectations that define a position. Epstein (1982) elaborates this connection between status and role. A woman's role is defined in society in the way she is expected to behave in certain situations, and her status as the esteem in which she is held by others who come in contact with her by virtue of her being a female. Valuation and definition of roles and positions in their several dimensions including cultural, biological, behavioural (interactive), socio-economic, etc., contribute to determining status in society.

Status amounts to how one perceives one's position of power, prestige and wealth, on the one hand, (i.e. self-worth) and how others perceive the position of that individual with regard to those constituents on which depends an individual's entitlements in society. One's self-worth, others' perception and entitlements come together to crystallise status. The social interaction in which status crystallisation of a person's position is arrived at depends also on his/her ascribed and achieved status. Status here becomes a position derived through an interactive process in the giving, receiving and maintenance of prestige, privileges and legal rights. Privileges are crystallised into a system of legal and economic immunities from external control, protected by custom, religion and law. In sum, there are objective criteria, such as economic resources, age, sex and ancestry (mostly ascribed), and power, prestige and wealth – which are achievable, but only through performing socially desirable and valued roles. Others reward one through these considerations that enhance one's status. Through the desirable obligations of an ascribed status, a person achieves enhancements of honour or prestige. In my ethnographic study (1994), a young bride or daughter-in-law enhances, by virtue of her arranged wedding, her own honour and that of her natal and conjugal families. If matched wisely as per social norms the two families gain social capital within an unequal share of the honour between the wife giving and the wife receiving families. It is this fluidity in an interactive process (along the lines acknowledging achieved and ascribed status) that calibrations in social relations of power, prestige and wealth continuously take place which constitutes status formation. Bourdieu (1986) has lent a greater fluidity and flexibility to the idea of reproduction of status and social distinctions. Status reproduction may be applicable to individuals as well as groups (Weber called them status communities or estates also). This paper will consider the aqueousness of social gauging for self-worth, entitlement, power, prestige and wealth as the 'status game' at work in individual and status groups.

Women Considered in Population Studies

The 'voluntary motherhood' demand by the women's movement was the first explicit demand for birth control in the 1840s. In 1914 Margaret Sanger's journal, 'The Woman Rebel' was founded. Birth control was sought in order for women to have rights over their bodies, reproduction and to be healthy. It was to enable women to escape out of one pregnancy after another.[1] Sex *sans* reproduction was no longer a sin. Shulasmith Firestone in the classical text, *The Dialectic of Sex* in 1974 advocated the elimination of female reproductive function to achieve equality with men (caricatured to show all radical feminism as essentialist or biologically determinist). The knowledge for control pertained to culture and thus to the male, science, and technological domain. The abortion struggle was for the right to privacy (to abort) in the U.S., and the right to women's health in Canada. The right to contraception was liberating to women, by giving them more control over their bodies and their reproduction, it enabled women to emulate men. By being like men, women would improve their position and aid in the overall development of society. Even with the pro-life and pro-choice strands, the above perspective was not lost sight of (See Jacobson 1990 for the controversial theoretical, theological and feminist views on abortion).

Population had gained prominence as a social problem with Malthus' essay on population. Subsequently, Malthus was often sought when feminists found his arguments suitable in their struggle for birth control. The more serious relationship between population and women emerged in the 1960s followed by the unsuccessful developmental path of population predicted by the theory of demographic transition. This theory interpreted the largely demographic (fertility and mortality) interplay hinged on urbanisation, industrial development and modernisation.[2] (Madhok in this volume with reference to the Princeton Project.) The question arose whether demographic knowledge was capable of designing the world's population. Other variables were sought to explain high fertility despite fall in mortality leading to population bulges. 'Culture' came handy as variable to serve the purpose of demography (Fricke 1997). Demography which had been sharing concepts with some other social sciences, opened its doors to anthropology through the use of 'culture'. Besides 'value of children', women as a category gained attention. Through a comparative understanding of world demographic trends the lesson was that if women from the developing countries were to become like those of the developed ones, they would achieve high status through education and paid employment and have far fewer children. In any case, for most measures of fertility women are central. Each fertility rate is worked out with women as the denominator. With attention to culture, they began to be studied in cultural settings and communities.

From Women to Women's Status

In the developed Western countries, women had to struggle against the eugenicist and classist ideologies of the population control establishment that wanted white, upper-class, educated women to have as many children as possible. These women desired the contrary. Women's fight was for birth control rather than population control (inspired by Malthusian ideology). Shapiro (1985) describes the controversies and politics of population control and birth control in the U.S. for over a century since the 1840s (this politics has not yet been resolved). Elsewhere, non-Western women's higher fertility, much against their governments' priority, pointed towards their low status. Women's status and role became significant as explanatory variables. Therefore, the argument arose that women with higher status prefer to have control over their own bodies and have fewer children whereas women of low status have no control over, their bodies have more children. This assumption ignores the role of socialisation and psychoanalytic views on reproduction and gender, where women are taught to be 'tongueless', 'desireless' and 'submissive nonpersons' (Bhatty 1988).

The expression, 'status of women' entered demography through two additional paths. The United Nations (UN) in the early 1970s initiated projects on women's status in several countries all over the world. The Committee on Status of Women in India, 1974 is an example. Status of women and demography was one of the areas of discussion for the women's decade and 1975 as the women's year. Second, the issues of reproduction, abortion and contraception that the feminist movement's second wave raised were considered by the UN, as well as demography. Within a few years, several studies on women's position and/or status in relation with population processes, particularly fertility were conducted. (See Dixon-Mueller 1993 for details of the first such study for the UN). Of course, foreign missions had entered developing countries followed by field surveys (see Madhok in this volume). These showed low knowledge of modern contraceptives and negative attitudes towards birth control. Large scale surveys wield 'bio-power'. However, by the 1970s the shaky grounds on which family planning activities were based began to surface and interdisciplinary research came into vogue.

By applying the Malthusian assumption that population growth increases poverty, neo-Malthusian analysis argues that pregnancy leads to suffering, powerlessness and low 'status' of women. Bandarage (1997: 52) refers to a chart prepared by the Population Crisis Committee ranking countries according to the status of women, showing a strong inverse correlation between women's status and fertility. The chart shows that pregnancy is the fundamental cause of women's poverty and powerlessness, thus fertility control the primary means of poverty alleviation and women's empowerment. Both large scale surveys in demography and smaller surveys following their pattern, influenced by the new home economics school, have used status of women as a variable. A major breakthrough came with Dyson and Moore's (1983) paper. This paper takes up some of the studies for further analysis. Numerous studies on the status of women's relationship with fertility took place in the 1970s and 1980s, until the early 1990s (Safilios-Rothschild 1982, Mason 1984, Basu 1992). They reiterated inverse correlations between women's status (based on objective socio-economic indicators) and their fertility worldwide.

Women's status was understood by most of the studies on fertility in terms of objective, quantifiable socio-economic and demographic indicators. These include duration of marriage, family structure, family income, economically active and dependent family members, caste, education and occupation of husband, property ownership, education and gainful employment of women (Mahadevan 1979, Singh 1979 among others). Social and cultural variables were additional ones in the studies by Chaudhary (1982), Dyson and Moore (1983), Mason (1984) and Basu (1992), for developing countries. Yadava (1995) uses the terms 'women's status' and their social status without clarifying how they are distinct. Perhaps decision making and interaction with the world outside the household is what Yadava implies in the term, 'social status'. These studies overlook status as a multitudinous construct, from the Weberian and Bourdieuist perspectives as aqueous and being constantly carved and recarved through everyday behaviour, economic as well as cultural, social and political influences. It combines system with process, a relationship of dependence with freedom continuously changing while remaining more or less integrated.

Differing Meanings of Status

Influenced largely by policy considerations, demographic studies on women, by and large, focused on the reproductive age group of fifteen to forty-five. Dyson and Moore's (1983) study went beyond this reproductive age group. They viewed fertility outcomes in the context of kinship, marriage and family practices as providing a certain kind of autonomy to women. It differed strikingly between North and South India and so did their fertility. Dyson and Moore (1983) showed a higher degree of gender stratification in the North Indian patriarchal culture than that prevailing in South (excluding East) India and this explained the fertility differential. The patrilineal kinship in North India is not as congenial for the autonomy of the female as the matrilineal kinship in South India. Their approach is further reflected in the subsequent study by Basu (1992). In spite of the North Indian kinship structure prevalent in the relatively developed Punjab, it is neither kinship nor religion but socioeconomic factors that explain, for Dyson and Moore, the higher status of women there. In other words, Punjabi women's comparatively low fertility within the North Indian region (by implication higher status) does not fit with its strong patriarchal kinship culture. Punjab is in this sense an outlier.

Middle East and North African kinship systems are similar to those in North India with predominance of female seclusion (female seclusion is lesser in North Africa). Kritz and Gurak (1989) show that sub-Saharan African women have their own plots of cultivable lands and engage in trade. They have considerable economic independence, access to resources, freedom of movement and decision making powers. By Dyson and Moore's (1983) logic, sub-Saharan African women's status should be high and their fertility should be lower, which is not really the case. In matters of marriage and reproduction these women have little decision making power. Yet again, it is relevant to explore why women in East Asia, South-East Asia and Latin America, who are in a more egalitarian relationship, rarely secluded and with near equal literacy and economic opportunity to men, find that son preference does not matter to them as it does in other societies. Closely related is, how in the course of economic development did Thai women gain an egalitarian relationship in matters of literacy, economic independence, financial control and also an absence of son preference (Knodel et al. 1987). Non-seclusion and economic opportunity (rather a significant cultural transformation owing to economic development) in Thailand resulted in increased status of women and control over their reproduction, while the same features despite being traditionally ingrained in sub-Saharan Africa did little to provide such control to the women there. These contrasting cases are little help in providing a handle to what actually raises women's status or conversely what retains gender inequality. Neither non-seclusion nor economic activity as variables are sufficient to explain status improvement or vice versa. How do we then decipher status?

Different definitions of the status of women imply a comparison with the status of men. Women's rights (Dixon 1975), women's status (Dixon 1978), patriarchy (Cain et al. 1979), men's situational advantage (Caldwell 1981), female power and autonomy (Safilios-Rothschild 1982) and female autonomy (Dyson and Moore 1983), are examples. Epstein (1982) focused on esteem accorded to women by virtue of their gender. This is at variance with Dyson and Moore's (1983) definition of status as attitudes towards women on the part of men. Women's access to and control over material and non-material resources determines their status for Cain et al. (1979) and Dixon (1978). For Dixon (1978) material resources mean food, income, land and other forms of wealth, and social resources consist of knowledge, power and prestige within the family, in the community, and in the society at large. Power and prestige (esteem) are more exclusive dimensions of status in comparison with access to and control over resources. Men's situational advantage (Caldwell 1981) and patriarchy (Cain et al. 1979) are seen as a set of social relations with a material, physical, emotional, social and cultural base that enables men to dominate women. Patriarchy describes the distribution of power and control of resources and women in which women become dependent on men and their resources. From the above descriptions of material and non-material resources unequal sexual relationships are evident. Men have and exercise power to control resources and women, despite differential access to economic independence and spatial mobility. But in all situations, women's adherence to family ideology through an acceptance of men's dominance seems like a hegemony of sorts, to use a Gramscian idea, in the context of the family and gender relations. The running of the family is the common goal for both men and women within the culturally acceptable role distribution and freedom. Then the question arises whether women's status is to be understood within the fold of such a hegemony or without it.

There are variations within the macro picture. Data from my village study in North India (Patel 1987) found that economic opportunity, wage labour, non-seclusion were culturally constructed, and differed with caste (higher or lower) membership. The honour accorded to women's work or seclusion is based on cultural meanings. As women advance in the household cycle, their status improves gradually. This is a kind of reward to those who conform and get rooted. 'The mother-in-law is most effective in channeling the institution of patriarchy in the household' (Patel 1987: 44). She and other older women enforce customs and are also responsible for socializing younger girls. Daughters-in-law are kept under control especially in matters of reproduction. Srinivas and Ramaswamy, based on several sociological village studies in India, attest to this fact that, '...the woman has control over the reproductive career of her daughter-in-law, but not her own' (1977: 13). Ironically her interests in these two roles are quite different. Status in the household may not always coincide with status in non-household contexts, such as in the city, in employment, or in the natal home. Status could vary slightly or largely on the position and role of an individual and her household, social, political context, as she shuttles from one to another context. Little attention has been paid to this dimension of status.

Age, number and sex composition of surviving children enhance a woman's status more than non-seclusion (freedom of movement) and economic opportunity. Whether they prefer their daughters-in-law to curtail reproduction or not has demographic significance, but what is crucial here is that age, seniority, motherhood and finally, mother-in-law hood as a composite of social and cultural factors cumulatively enhance women's status (Patel 1994). Even women's animal-like existence described by Mayo (1927) attests to the immense authority of the mother-in-law over the daughter-in-law. This however happens only after the mother-in-law is past her prime. It is worthwhile to ask if the sub-Saharan African women ever obtain decision-making powers regarding their reproduction or that of their younger generation females. In many other cultures where women are in an egalitarian relationship, older women gain considerable decision-making authority, control over resources and some degree of freedom from seclusion. Further, is it possible to infer, as is done in demography, that independence, mobility and economic activity imply the higher status of women? The contrast between the dread which Indian women have of a second wife being brought in by the husband and the sub-Saharan African women's preference for a co-wife (Kritz and Gurak 1989, Van de Walle 1993, and other anthropological evidence) is difficult to explain unless local cultural meanings are made clear. For an Indian woman a co-wife amounts to her loss of social, economic and emotional status within and outside the household, while it works the other way round for the sub-Saharan African woman. She gains powers over a young co-wife, similar to a mother-in-law's gains in North India. Also, this attitude towards the co-wife is not the same all over Africa. Unlike the Sahelians who take great pains to keep the harmony of the compound, among the Yoruba the word for co-wife means 'the jealous one' (Van de Walle 1993). Variations in status across cultures, age, class, etc., are enormously complex. What is prestigious and power giving in one culture is not the same in another. However, fertility and childbirth are crucial for identity and status in all cultures. The variation, however, is regarding the number and sex of the surviving children, and ideas about by what age to have babies. Such cross-cultural variation does not prove helpful in objective measurement of socio-economic indicators of status that are useful for demographic calculations.

Even within a given caste, a community, each status group or community has a similar life style and common moral system. The social functions and history produce separate, united communities whereby, for instance, daughters-in-law and mothers-in-law constitute separate status groups, with differing political and legal entitlements, in the same society, despite being from the same households or families, and with the same economic standing. In this regard, the literature that excludes the inequality and power relations between women, for example Mason (1984, for valid reasons) misses out the enormous evidence in demography, sociology and women's studies that a dissimilarity in status exists among women positioned differently, even in the same family.

Women's role obligation and the meaning assigned to it is a major determinant of women's status in society. This is so even in Pakistani society (Sathar et al. 1988). The women are viewed primarily as wives and mothers and they must marry and reproduce to earn status. Pakistani women are similar to sub-Saharan African women in terms of decision-making ability, but differ from them in their disability to engage in gainful employment. On the basis of examining the two conventional measures of women's status, namely educational attainment and labour force participation, Sathar et al. (1988) contend the association between higher levels of education of mothers and lower level of fertility. It is difficult to assess whether it means a stronger relative position of women within the household and greater voice in family decision-making. The authors are fully aware that female education can affect fertility by several routes which may not be indicative of women's status. The role of education in reproducing the dominant class culture and gender ideology opens a critical window as to how and to whom education serves (Bourdieu and Passeron 1977).

Little or no association was found between work and fertility in Pakistan. This is because women may not work out of choice and would give up work as soon as their financial condition improved. Not all work by women is acceptable let alone valued in the society and the group. Respectability and enhanced status is associated only with teaching and medical professions. Other jobs, though they accrue income and enable non-seclusion, are not seen as respectable (cf. Patel 1987 for rural North India). Numerous studies since Boserup's (1970) pioneering work have revealed how women's integration into the development process writ large has served to deepen their subordination. In the Indian context, they become more sanskritised and consequently subordinate to their betters. Bardhan (1985) demonstrated how developmental learning in poultry, gardening, cookery, etc., did not necessarily lead to improved status of women in the household. Sociological and social anthropological evidence is that work is about control, evaluation and identity (Wallman 1979). Godelier (1972) states that unlike economics where work is for maximisation of capital, it is an economic, political and ritual act, all at the same time. Economic activities, social and ritual ones are tied up with fertility and gender relations. Sharma and Vanjani (1993) also corroborate this for rural Rajasthan. The value of work depends on where in the hierarchy, in any given society, does the actor sit.

Lack of communication between spouses is reported to enhance fertility and minimise contraception. Poffenberger (1969) has elaborately described the lack of communication between spouses in Gujarat (India). Lack of communication is most prevalent among young spouses living in joint households, while those in nuclear households communicate better and have fewer children. But my data refutes this thesis on more than one count. In rural Rajasthan residence after marriage is patri-virilocal and the newly married couple is obliged to live as part of a joint family. This period happens to coincide with their initiation into their fertility career. It is their fertility performance more than any thing else that makes their marriage viable and the future course of their family life feasible. Interspousal communication is no substitute for having babies. To have two to three children before the couple secedes from the joint household to set up a nuclear one is common. Modesty obligation is also found in Sharada's (1977) study in South India. Little communication or discussion happens between spouses. 'Even where women were assertive decision makers in family matters, they were reluctant to speak about sex and fertility-related issues with their husbands. Many women felt inhibited and thought it unwifely to do so' (Sharada 1997: 121).

The image and value of motherhood in society earns social honour and prestige for mothers. Though, fatherhood also adds to one's esteem in social reckoning, the woman's gain is more visible. Mayo (1927) was struck by the honour a woman gained within the *zenana* (women's quarters) upon the birth of a son. Fertility and motherhood come together to create and maintain power forms through political monopoly, cultural reproduction and social exclusion (cf. Inhorn 1996 for how dishonouring and agonising it is to be a childless woman in Egypt). The interactive processes and protection of certain privileges and honours goes on. Prize-giving, symbolic of granting public honour, contributes to a group's integration and identity, and these reaffirm commitment to group values within the norms of fairness and etiquette. The husband and wife and daughter-in-law and mother-in-law interests may clash, and conflict does occur in the interactive process while exclusion from privileges and the award of public honour also do take place around socially valued performances and achievements. A young bride in my study is provided some privileges and protections to express inclusion while she faces relatively stricter seclusion. But her honour enhances through the public acknowledgement of her motherhood. Most women consider motherhood as an essential element defining their lives and proving their worth to themselves and to others. It is easier to desert a wife but not easy to desert the mother of your children. The most poignant relationship exists between mother and son. It forms a nexus. The fear of her husband deserting her and/or bringing a second wife is wiped out. She earns her right to live in the family through her son's birth. Being a son's mother is a license to be a potential mother-in-law that holds more exclusive privileges to women. It is a sure source of status for the mother.

Through individual quest for status enhancement, the status of the group, such as the household, the family, etc., also increases. Each daughter-in-law in her attempts at motherhood, not only gains rewards, prizes and individual privileges, but also attains distinction (Bourdieu 1986) by entering into the category of mothers as a status group. In comparison with economic classes, mothers as status groups are characteristically social collectivities of a communal nature with conventional social reproduction and cultural inheritance. Prestige is a social-psychological category and claims to prestige are recognised by others willing to give deference. On awareness of prestige ranking depends status. A distinct lifestyle confers a special identity in society. An old man in my study (Patel 1994) married three women one after the other in the hope of attaining fatherhood, especially by having a son, as he lost out on the honour others of his age group had through children. The relief from arduous chores and by earning a living that grown up children bring to their middle-aged parents brings them prestige and honour in society. Their absence amounts to some kind of destitution, both for the elderly male (here the interests of the spouses are common), as neighbours and relatives are reluctant to offer support when reciprocity is unlikely. Being deprived of company, security and physical and emotional support in the absence of children fails to enhance social status, if anything it wanes status. The nexus of status and fertility is multiplex.

The idea of the individual in Indian society is less important, except at the existential level. The society is as it were, characterised by collectivity, while the individual is particular to the West. The idea of individual status derived from objective indicators sieves through the existence of the individual as part of the overarching collectivity. It is the family and the household that has overarching priority for an individual, more so a woman. The cultural definition of the female and her identity is largely in association with the inside, the home and the courtyard, where she cares for the family. Even when a woman is outside she feels her interest should be inside (see Unnithan-Kumar 2001 for significance of collectivity and emotions in women's reproductive agency). Sharada observes for south India, 'It is not just psychological or geographical but it is also social, in the sense that you should be oriented to the family all the time, whether you are pursuing your career or whether you are staying at home' (1997: 125). Rao resonates for the Himalayan nomadic herders. Even when women in nuclear families have greater 'autonomy' in receiving or visiting natal kin than in joint families, these are only matters of rejoice. 'This greater female autonomy is however not seen as authorising individual interests which could counter those of the household as a whole, since the well-being of the individual is normatively held to be inseparable from that of the household. To take a decision that runs counter to the well-being of one's household is considered foolhardy if not outright abnormal and even immoral' (1998: 237). It is not only for the oriental societies that the concept of status derived from the western notion of the individual poses limitations. Chodorow (1978) and Gilligan (1982) show the proclivity to place a premium on attachment, care, love among females as against separateness and competition among males on the basis of their studies in western societies.

Assumptions of Status: Possibility of Lessons

As we have seen above, not only does status vary cross culturally, it is fluid within cultures and even for an individual during their life cycle. Status is too fluid a category to be given the abstracted objective, scientific treatment, a prerequisite in demography. The kinship contrast between matrilineal and patrilineal societies on the one hand, and on the other, the several economic and developmental dimensions such as seclusion, employment and literacy of women fall short of a standardised scale to grade societies on. Status does not incorporate certain universal attributes across societies to arrive at a general yardstick for fertility outcomes. Women are differentiated either/and by age, cohort, economic background, caste, motherhood and fertility within the same societies and they differ across societies. The other difficulty is the assumption that women's status is not an isolated and independent variable.

Women are graced with status through socio-economic development. In demographic calculations, aggregates of education and employment of individual women correlate inversely with fertility. Hence, the calculation that education and employment are status enhancing. Studies on the relationship of education and fertility provide conflicting evidence. On the other hand, there are studies, such as Mason's (1984) which do not see the impact of fertility on women's status. She seems to trace the possible causal paths between aspects of female status and those of fertility and mortality and not look for determinants of status and by implication, of agency. The stress is to focus on status of women in comparison to men. Thus the logical corollary is that when women act like men (through education and employment) their status would rise. But what is crucial is to see the pathways to status as much as it is to see those to a decline in fertility. We need to pause here and look for any Eurocentric assumptions in the pathways. The objectivist and instrumentalist models of social relations, in reality, are constructed somewhere between institutional dynamics on the one hand and social as well the self-referential on the other. We should ask if research is considerate enough to see the complexity besides the variability.

As status suffers from the limitation, especially in its inability to provide causal pathways to fertility, female autonomy is used to imply more freedom and power in decision making. But a high status may not necessarily lead to autonomy. Status may be lost by earning an income if female seclusion is highly valued and expected. It is the social cost which is paid for by the economic returns from female work. To quote Jejeebhoy, 'What is important for demographic change is that women be in control of their own lives and have a voice in matters affecting themselves and their families, not how much prestige or esteem they are accorded' (1995: 7). To quote Jeffery and Basu (1996: 16) 'the use of status has been heavily criticized for appearing to focus on the division of labour and for failing to deal adequately with issues of power and conflict. The ability to manipulate their environment accrues to many women from the developing countries through their reproductive performance. Subsequently, other sources may include literacy, employment, interspouse communication, etc. How has the term autonomy evolved in usage in relation to women and why has it replaced other terms such as women's status? To understand this both the terms have to be critically analysed. Status and autonomy are both affected by attitudes, values and customs besides legal traditions. (For a logical analytical discussion on autonomy, see Madhok in this volume.)

The difficulty with status is that it does not mean power, which refers to women's ability to influence and control at the interpersonal level. It is the multidimensionality of the concept that poses problems for demographic correlations. Mason (1984: 8–7) elaborates this point rather succinctly. Following this logic, Mason finds Safilios-Rothschild (1980) and Youssef (1982) as making contradictory points, which is not really the case. The latter emphasises that non-seclusion and economic independence of poor women go together, while the latter observes that the family sacrifices social status it could have derived from its women's seclusion. The two studies focus on women's economic activity but with different analytical tools. The latter sees the association while the latter sees the meanings of such an association. Patel (1987, 1994) and Sathar et al. (1988) discussed above, corroborate this view, as does Schildkrout (1979) below. The case of sub-Saharan African women's active economic participation mentioned earlier, challenges this assumed causal link and correlation. Because status is not just an individual attribute, it is interactive with the family in that it is received from and extended to one's household and family. The case of birth of male children is not one of sacrificing individual over collective interest, it very much serves individual interest as well. Caldwell (1999) in his study on marriage in Sri Lanka, refers to religious duties among Hindus, which are strictly speaking, not necessarily for individuals. They are as much for the family. Individual desire is the source of evil and is derived from self-centred actions.

My studies (Patel 1994, 1999 and 2002) give evidence of women arresting reproduction after attaining the 'social optimum'. Respondents tend to acquiesce and give responses they think researchers might like to hear (Mamdani's 1972 respondents accepted contraceptive pills for the sake of politeness and to help researchers gain promotions!). Women in my study desired to arrest fertility when they were cursing children or lost patience over something. But when it came to serious decisions toward contraception, they had several fears. Those who had lost no child, and nor had their kinsmen, were keen to try contraception. The tenuous character of the statement needs to be unravelled rather than taken literally. Nevertheless, it is important to note that this is a juncture in women's lives when, *pari passu*, they desire to curtail reproduction. And it is not low status or lack of autonomy alone but a whole range of relationships and factors that come into contemplation and its logical culmination (see my work, Patel 1994 for the decision-making process). The idea of the social optimum can be seen in association with that of the threshold, both individual and social. The concept, 'unwanted fertility' adds another dimension to the complexity at hand.[3] Fertility needs to be situated (Greenhalgh 1995).

An assessment of female status on any single dimension may be too partial and ephemeral. Her status depends on the simultaneous and multilocal contextual factors (Mukhopadhyay and Higgins 1988, Patel 1999). Moreover, various components of status may move in different directions at a given period of time, so that defining what constitutes 'improvement' is a complex proposition. Besides women's education and work, dowry is another indicator. As mentioned earlier, dowry is seen as a symbol of high caste status and status mobility, while it actually intensifies discriminatory attitudes towards girls, and also negatively affects their perceptions of self-worth. But at the same time, bride price has come to be associated with low caste status. Studies of working poor women, heading households all over the world provide evidence of women spending their income on the household and the children rather than on themselves (see Patel 2001 for references to Chant and Gulati among others in this regard). Schildkrout (1979) reports about Muslims in Kano city in Nigeria where girls' income is invested right back into the marriage system, i.e., for dowry and related marriage expenses to raise the status of the girl and her family. Early marriage, high fertility, submission to male authority in decision making in domestic and political domains, polygamy and seclusion or purdah in marriage are crucial for women's status, even though the women from Kano are economically not dependent on husbands and sons. Also, Western education among a section of the Hausa is a means to find an educated match rather than for better economic opportunities as these women are expected not to work after marriage. Thus education may have a value in itself and be a source of status enhancement for educated women, but not because it accrues income. It is through the ability to associate with socially desirable males that women gain status, rather than being independent or having personal interests in competition and conflict with their men. It is through association with the socially desirable and prestigious that one's honour enhances. In most societies women enjoy power because of who their fathers and husbands are. Women's status is dependent on men even when women are not economically highly dependent on men. Dube (1988) provides the use of positive terms like *'saubhagyvati'* or *'suhagan'* for women whose husbands are alive, while avoidance of widows and barren women on auspicious occasions. When status is derived through association with males, it introduces complexity in an individual's status as an isolated category.

Thus assessing status involves a socially considered judgment of what is good, desirable and valuable. Such criteria are more often than not, culturally specific. With a large number of possible status indicators and the presence of subjectivity in assigning 'high' or 'low' status to a given indicator, it is difficult to examine comparatively how women are placed or positioned in different sociocultural milieux. In the absence of a unitary phenomenon to be cross-culturally measured and compared to study gender asymmetry, the use of the term autonomy came into vogue. The autonomous capacity of the woman as an individual is a positionality of differential marking of aggregates of individuals. The very notion of autonomy, self-produced and self-identical, like that of a self-sufficient economy or absolutely sovereign polity, is constructed through demographic assumptions. The discourse has as its counterpoint the 'Other', usually the spouse, the mother-in-law and the culture. Though the 'Other' is different, there is a 'difference' as well, which need not be overlooked. How do women themselves perceive 'autonomy'?

During the 1990s, particularly the build up for ICPD (International Conference on Population and Development) in 1994 in Cairo and since then, the preference is for the term 'empowerment' of women over autonomy. The implied tension in using autonomy, which was always vis-à-vis some other individual (mostly male/spouse in the context of fertility), is overcome by the use of the expression 'empowerment of women'. But the notions of identity, self-worth, belongingness, honour and prestige, all influential in the making of status continue to remain central to societies and particularly to women in the context of reproduction. It is therefore important that the advantages derived through studies of women's status and autonomy as well as the limitations that have come to light are not thrown away with the new bath water.

Conclusion

The demographic individual is embedded in the twin notions that the complex factors accounting for human reproductive behaviour can be first isolated and measured and then manipulated. Accordingly, individual status is the subject. Individual women form the centre of demographic deliberations. The concept of status in reality is inherently a socially situated one, whether it deals with an individual, a group or cohort. Aggregates of individuals make enormous statistical sense, but their intricately tied roles and values complicate matters. An individual is related through kinship and community association with the larger group in society as is his/her esteem. Thus status is a fluid category both for the individual and the group. It also varies over the life cycle. The sensitivity required to understand the multilocal status constituents is quite difficult with standardised questions and desired cross-cultural generalisations. Like large scale surveys, questions on status are tested for their workability rather than their suitability. The latter is possible only when local contexts and the respondents' meaning systems become familiar. At this point the conventional survey questions are threatened with human subjectivity and researcher's bias, to remain objective. Even within a society, an objective abstraction and/or aggregation are not easy with respect to status.

Survey questions also assume the autonomy of an individual in questions pertaining to perception, decision and association. But even before the questions of perception arise, the question of meaning of the concept about which perceptions are to be studied needs to be explored. The break up of the concept into its constituent elements and factors need to be chalked out. The values and conventions of the concept are to be gauged. Thus status of women would require the incorporation of the locally perceived desirable behavioural and situational conditions, the factors and the dimensions that constitute status in society. Questions about what is desirable status, what enhances and changes it, may then follow.

Some of these problems remain in surveys because analysts are not the same as data collectors. Also, the hierarchy between the two is such that data collectors are rarely encouraged to express and seldom thought capable of new and different views and ideas about research. The distance of the analysts from the ground reality perpetuates the gap between what people think and what surveys show. Human actors live out their lives within a taken-for-granted reality that they construct collectively through their social life. To know what people will do next and how they will behave, it is necessary to comprehend their world. The analyst has two sets of things to understand: the world created by the actors and only then the world created by the analyst. The latter must be built on the former if it is to make any sense, which is the double hermeneutic. A word of caution is in place here not only for the demographer but also for the social scientist in general. Since the analysts' conclusions are parasitical on the actors' framing of their worlds, and since social lives are bound in time and space, there can never be a universal predictive social science in a way that there can be a natural science. When demography ventures into sociological and social anthropological arenas of status, and cultivation of social and symbolic differences, its ability at prediction is severely limited.

The relationship between association and explanation is to be clarified. Surveys find large data sets wherein variables show associations with different degrees of statistical significance. Large data sets increase the degree of significance. But association may not be an explanation. For instance, education and age at marriage may be positively associated but increase in education does not directly lead to increased age at marriage which in turn is inferred as improvement in status. In Kano, Sri Lanka and India mentioned above, the amount of dowry to be given by the bride's parents and other expenses incurred upon marriage increase as her education increases. It thus takes longer to accumulate the requisite amount, thus the delay. In the above association, the explanations are derived from an understanding of people's worlds. The explanations also require an insight into what are the aspirations and values for which education is a means.

The classical sociological perspectives on stratification are traceable in the manner in which the concept of status is used in demography. More common place is the socio-economic approach to women's status shown through education, labour force participation, access to income and its control and raised age at marriage. This approach is closer to the Marxist view on inequality dependent on the economic factors. As seen above, a few studies go beyond the economic criteria to incorporate the cultural and social ones in studying status of women. Patriarchy, men's situational advantage, culture, kinship, seclusion and esteem are seen to form status. This is closer to the Weberian perspective on status. But when attempting to impact the concept of status the interpretative accent of Weber gets a short shrift. The need however is not to discard the social and cultural location specific temporalities but to consider them central in making the status of a person in a society at any given time. The problem is rather serious in terms of how to view commonalities and what to do with differences. The question gets more complex when it is to be applied to fertility behaviour, which constitutes the challenge for demography.

Notes

1. Manju Kapoor's Difficult Daughters narrates the story of how the protagonist's mother, Kasturi had to be moved to the hills from Amritsar (Punjab in India) after the delivery of her eleventh child to rejuvenate her emaciated body and regain health. (The unsuccessful attempt at averting the eleventh foetus had made her and the *dai* [traditional birth attendant] reconciled to fate.) The family business of a jewellery shop enabled the luxury of a cottage on a hill station unavailable, to a very large majority of Indian women. Thus Margaret Sanger's observation that in the 1920s she found a large number of Indian women expressing their desire to arrest reproduction and seeking birth control means from her (India office library, London). Indian women were also participating at the wider level. Neo-Malthusian League's Indian branch was active and so was the leadership of the AIWC (All India Women's Conference). However, unlike the Neo-Malthusian League's views on birth control for the poor, Kamladevi Chattopadhyay of AIWC, had in those days, advocated birth control for women's rights (see Anandhi 1998).
2. Given the spread of capitalism in the world after the Second World War, the economic disparity among countries could not be factored in as the cause of high fertility and altered with remedial dosages of economic correction. Instead the ideology of interventions within the capitalist state was seen as more suitable. Population control could be achieved by altering women's attitudes, especially through education and health. The IEC (information,

education and communication) programme is a means to create favourable attitudes towards state population prescriptions.
3. The use of the term 'unwanted fertility' should be traced historically. Since the mid-nineteenth century, middle-class American women have wanted to keep reproduction within their control and preferred it to be voluntary. It was in the late 1960s that the term 'unwanted fertility' was used in population control circles to reassure people, especially the Right, that the state would not regulate reproduction. The programme instead promised social stability. The right to choose the elimination of 'unwanted fertility' became the chief selling point of the American welfare state (Shapiro 1985: 76).

There is also a need to distinguish want from desire. When a woman who reports to surveyors that she does not want any more children or replies in negative to their question, if she wanted more children, it has methodological as well as semantic implications. Methodologically, there is an acquiescence effect that engenders her to stating that she does not want any more children. Of course, she would not say this unless she already has what she thinks is the optimum number and sex composition of children. Recall how Mamdani's (1972) respondents reported that they had accepted contraceptives from researchers in the Khanna study for several years without using them much as it was impolite otherwise, and acceptance helped researchers move up in their jobs. Secondly, desire is different from want. A woman with an optimum number of children and when tired of arduous routine chores is more likely to desire an altered situation. She may not actually want to stop her reproduction when she expresses her desire to have no more babies. Nevertheless, she no longer desires to have more children (see Sharada 1997 corroborating that after a certain parity the desire for children was reduced). The semantics of the expression have to be contextually fathomed. Besides, women throughout history are known to have adopted various (even secretive) ways to avoid reproducing, though these have not always been easy or fruitful.

I am grateful to Profs. Mary Searle-Chatterjee and Ursula Sharma for their perceptive comments on this paper. Thanks are due to Dr Maya Unnithan-Kumar for holding the volume back to get my paper in and being patient with the pace of its progress.

References

Anandhi, S. 1998. 'Reproductive Bodies and Regulated Sexuality: birth Control Debates in early twentieth century Tamil Nadu', in M. E. John and J. Nair, eds. *A Question of Silence*. Delhi, Kali for Women, and London: Zed Books, 139–66

Bandarage, A. 1997. *Women, Population and Global Crisis*. London: Zed Books

Bardhan, K. 1985. 'Women's Work, Welfare and Status: Forms of Tradition and Change in India'. *Economic and Political Weekly*, 15, 51–56: WS72–8

Basu, A. M. 1992. *Culture, the Status of Women, and Demographic Behaviour: Illustrated with the Case of India*. Oxford: Clarendon Press

Bhatty, Z. 1988. 'Socialisation of the Female Muslim Child in Uttar Pradesh', in K. Chanana, ed. *Socialisation, Education and Women: Explorations in Gender Identity*. New Delhi: Nehru Memorial Museum and Library, Orient Longman, 31–39

Boserup, E. 1970. *The Role of Women in Economic Development*, New York: St. Martin's Press
Bourdieu, P. and Passeron, J. C. 1977. *Reproduction in Education, Culture and Society*. London: Sage
Bourdieu, P. 1986. *Distinction: A Social Critique of Judgment of Taste*. London: Routledge and Kegan Paul
Cain, M., Khanam, S. K., and Nahar, S. 1979. 'Class, Patriarchy and the Structure of women's Work in Rural Bangladesh'. *Population and Development Review*, 5: 3, 405–38
Caldwell, B. 1999. *Marriage in Srilanka: A Century of Change*, Delhi, Hindustan.
Caldwell, J. C. 1981. 'The Mechanisms of Demographic Change in Historical Perspective'. *Population Studies*, 35: 1, 5–27
Chaudhary, R. H. 1982. *Social Aspects of Fertility With special Reference to Developing Countries*. Delhi: Vikas
Chodorow, N. 1978. *The Reproduction of Mothering: Psychoanalysis and the Sociology of Gender*. Berkeley, University of California Press
Dixon, R. B. 1975. *Women's Rights and Fertility, Reports on Population/Family Planning*. New York: The Population Council
—— 1978. *Rural Women at Work: Strategies for Development in South Asia*. Baltimore: Johns Hopkins University Press
Dixon-Mueller, R. 1993. *Population Policy and Women's Rights: Transforming Reproductive Choice*. London: Praeger
Dube, L. 1988. 'On the Construction of Gender: Hindu Girls in Patrilineal India'. *Economic and Political weekly*, 23, 18: WS 11–19
Dyson, T. and Moore, M. 1983. 'On Kinship Structure, Female Autonomy and Demographic Behaviour in India'. *Population and Development Review*, 9: 1, 35–60
Epstein, C. T. 1982. 'A Social Anthropological Approach to Women's Roles and Status in Developing Countries: The Domestic Cycle', in R. Anker, M. Buvinic and N. H. Youssef, eds. *Women's Roles and Population Trends in the Third World*. London: Croom Helm, 151–70
Firestone, S. 1974. *The Dialectic of Sex*. New York: Morrow
Fricke, T. 1997. 'The uses of culture in demographic research: continuing space for community studies'. *Population and Development Review*, 23: 4, 827–32
Gilligan, C. 1982. *In a Different Voice: Psychological Theory and Women's Development*, Cambridge, Mass.: Harvard University Press
Godelier, M. 1972. *Rationality and Irrationality in Economics*. London: New Left Books
Greenhalgh, S. 1995. *Situating Fertility: Anthropology and Demographic Inquiry*. Cambridge: Cambridge University Press
Inhorn, M. 1996. *Infertility and Patriarchy: The Cultural Politics of Gender and Family Life*. Philadelphia: Philadelphia University Press
Jacobson, J. 1990. *The Global Politics of Abortion*. Worldwatch Paper 97, Washington D.C.: Worldwatch Institute
Jefferey, P. and Jefferey, R. 1996. 'What's the Benefit of Being Educated: Girls' Schooling, Women's Autonomy and Fertility in Bijnor', in R. Jefferey and A. M. Basu, eds. *Girls Schooling, Women's Autonomy and Fertility Change in South Asia*. New Delhi: Sage, 150–83
Jeffery, R. and Basu, A. M., eds. 1996. *Girls' Schooling, Women's Autonomy and Fertility Change in South Asia*. New Delhi: Sage

Jejeebhoy, S. 1995. *Women's Education, Autonomy and Reproductive Behaviour: Experience from Developing Countries*. Oxford: Clarendon Press
Kapur, M. 1998. *Difficult Daughters*. New Delhi: Penguin
Krtiz, M. M. and Gurak, D. T. 1989. 'Women's Status, Education and Family Formation in Sub-Saharan Africa'. *International Family Planning Perspectives*, 15: 3, 100–05
Knodel, J., Chamratrithirong, A. and Debavalya, N. 1987. *Thailand's Reproductive Revolution*. Madison. WI: University of Wisconsin Press
Mahadevan, K. 1979. *Sociology of Fertility: Determinants of Differentials in South India*. New Delhi: Sterling Publishers
Mamdani, M. 1972. *The Myth of Population Control*. New York: Monthly Review Press
Mason, K. O. 1984. *The Status of Women: a Review of its Relationships to Fertility and Mortality*. New York: The Rockefeller Foundation
Mayo, K. 1927. *Mother India*. Bombay: Allied
Mukhopadhyay, C. C. and Higgins, P. 1988. 'Anthropological Studies of Women's Status Revisited, 1977–1987'. *Annual Review of Anthropology*, 17, 461–95
Patel, T. 1987. 'Women's Work and their Status: Dialectics of Subordination and Assertion'. *Social Action*, 37: 2, 126–49
—— 1994. *Fertility Behaviour: Population and Society in a Rajasthan Village*. Delhi: Oxford University Press
—— 1999. 'The Precious Few: Women's Agency, Household Progression and Fertility in a Rajasthan Village'. *Journal of Comparative Family Studies*, 30: 4, 429–51
—— 2001. 'Women and Migration', in S. Vishwanathan, ed. *Structure and Transformation*. Delhi: Oxford University Press, 131–51
—— 2002. 'Indigenous Fertility control', in R. Potter and V. Desai, eds. *Arnold Companion to Development Studies*. London: Arnold
Poffenberger, T. 1969. *Husband-wife Communication and Motivational Aspects of Population control in an Indian Village*. New Delhi: Central Family Planning Monograph series, No. 10
Rao, A. 1998. *Autonomy: Life Cycle, Gender and Status among Himalayan Pastoralists*. Oxford: Berghahn Books
Safilios-Rothschild, C. 1980. 'A Class and Sex Stratification Theoretical model and its Relevance for Fertility Trends in the Developing world', in C. Holn and R. Amchensen, eds. *Determinants of Fertility Trends: theories Re-examined*. Leige, Ordina editions: 189–202
—— 1982. 'Female Power, Autonomy and Demographic Change in the Third World', in R. Anker, M. Buvinic and N. H. Youssef, eds. *Women's Roles and Population Trends in the Third World*. London, Croom Helm, 117–32
Sathar, Z., Crooke, N., Callum, C. and Kazi, S. 1988. 'Women's Status and Fertility Change in Pakistan'. *Population and Development Review* 14: 3, 415–32
Schildkrout, E. 1979. 'Women's Work and Children's Work: Variations among Muslims in Kano', in S. Wallman, ed. *Social Anthropology of Work*. London: Academic Press, 69–86
Shapiro, T. M. 1985. *Population Control Politics: women, sterilization, and reproductive choice*. Philadelphia: Temple University Press
Sharada, A. L. 1997. 'Women, Fertility and Empowerment', in H. Afshar and F. Alikhan, eds. *Empowering Women: Illustrations from the Third World*. London: Macmillan, 115–130

Sharma, M. and Vanjani, U. 1993. 'Engendering Reproduction: The Political Economy of Reproductive Activities in a Rajasthan Village', in A. C. Clarke, ed. *Gender and Political Economy: Explorations of South Asian Systems*. Delhi: Oxford University Press, 24–65

Singh, K. P. 1979. *Status of Women and Population Growth in India*. Delhi: Manohar

Srinivas, M. N. and Ramaswamy, E. A. 1977. *Culture and Human Fertility in India*. New Delhi: Oxford University Press

Unnithan-Kumar, M. 2001. 'Emotion, Agency and Access to Healthcare: Women's Experiences of Reproduction in Jaipur', in S.Tremayne, ed. *Managing Reproductive Life: Cross cultural themes in fertility and sexuality*. Oxford: Berghahn, 27–52

Van de Walle, F., Van de Walle, E. 1993. 'Urban Women's Autonomy and Natural Fertility in the Sahel region of Africa', in N. Federici, K. Mason and S. Sogner, eds. *Women's Position and Demographic Change*. Oxford: Clarendon Press, 61–79

Wallman, S. 1979. 'Introduction', in S. Wallman, ed. *Social Anthropology of Work*. 1–24

Yadava, K. N. S. 1995. *Status and fertility of Women in Rural India*. Delhi: Manak

Youssef, N. H. 1982. 'The Interrelationship Between the Division of Labour in the household, women's roles and their Impact on Fertility', in R. Anker, M. Buvinic and N. H. Youssef, eds. *Women's Roles and Population Trends*. London: Croom Helm, 173–201

CHAPTER 11

HETERONOMOUS WOMEN?
HIDDEN ASSUMPTIONS IN THE DEMOGRAPHY OF WOMEN[1]

Sumi Madhok

Introduction

This paper is divided into two parts. The first argues that the conventional language of autonomy is unsuitable for illuminating aspects of social reality, particularly agency of persons within oppressive social contexts. The second part provides illustrations in support of this argument. This chapter suggests that a failure to engage with philosophical concepts in the light of different historical contexts can lead to not only an uninteresting analysis of social behaviour but also a potentially misleading one. I analyse the literature on fertility or what has come to be known as the 'demography of women' and its employment of a conception of autonomy. I argue that the understanding of autonomy within this literature is not only oblivious to feminist concerns about autonomy but also fails to pay attention to women's agency. The chief reasons for this failure lies in the conceptual homology fertility studies establish between individualism and autonomy.

Part I

Autonomy within Conditions of Subordination

How to think of autonomy within conditions of subordination? Furthermore, why do we need to configure autonomy within these conditions? The question at face value is inherently paradoxical. How can autonomy, which in some senses implies freedom, be understood within conditions of subordination? Or further, are not all of our lives constrained in some way or the other? Subordination is not treated here as a relative experience of disadvantage but refers to those constraints, both moral and social, which affect our abilities to put into effect our chosen desires. Perhaps the most important reason for reconfiguring autonomy within subordinate conditions is that subordinate circumstances characterise many women's lives.[2] In the light of a recognition of women's subordination, it becomes imperative that we tailor our conceptual tools in order to accommodate possibilities within which women express certain desires and choices either in contravention of their expected social or moral roles or where they are unable to effect these in their actions. Secondly, it is important to recognise certain agentic capacities within subordinate circumstances in order to avoid the construction of a victims discourse. To point to the subordinate circumstances of their lives is not to point to their lack of agentic capacities or the passivity of their existence. Third, subordinate circumstances are an aspect of the sociality of our lives. All circumstances of our lives are not autonomy enhancing, i.e., are not conducive to free choice. Paying attention to the sociality of our lives, our shared norms and meanings must also include the oppressiveness of our lives. Conceptualising autonomy within conditions of subordination pays attention to the oppressive social contexts within which many women live. There must then be an account of not only the everyday aspect of their lives, within which they display their agentic capacities. This account cannot be only one of freely chosen action.

The feminist philosophical interventions and their subsequent critiques of autonomy[3] have facilitated the philosophical possibility of autonomy within conditions of subordination. Feminist philosophers have expressed reservations on the conventional understanding of autonomous agents as 'self sufficient, independent and self reliant' (Code 1987: 358). Much of the feminist critique has emanated from the excessive individualism informing this conception of autonomy with its emphasis on distinctness and separateness of the subject. The critique of the decontexualised self interested autonomous individual has resulted in a project of reconceiving autonomy in ways that are receptive to feelings, emotions and to other relational aspects of human behaviour. This philosophical endeavour of reconciling agency with relationality has resulted in what is known as 'relational autonomy'. Relational accounts of autonomy are of two kinds: procedural and substantive. According to procedural accounts, individuals realise their autonomy by engaging in a certain kind of critical self-understanding and by following decision making procedures that are free from coercion and manipulation.[4] Autonomy within procedural accounts,[5] therefore, does not require that persons exercise particular kinds of choices or uphold certain kinds of normative competence. Substantive accounts on the other hand insist that persons uphold a particular content in their actions.

While the emergence of procedural and relational accounts of autonomy makes it possible to envisage autonomy of persons within oppressive contexts, this is in itself inadequate to explain autonomy within subordinate circumstances. The principle obstacle in these accounts, one that they share with conventional understandings of autonomy, is in their emphasis on action. Relational accounts, like nearly all accounts of autonomy, coalesce on the belief that an agent must be able to act on her beliefs. However, understanding the subordination of people's lives implies that we recognise that persons cannot always act according to their desired preference. Therefore, within these conditions, it becomes imperative that we shift our gaze from the final acts of persons to an understanding of the ideas that lie behind action itself. Looking behind action focuses on the process through which action takes place and uncovers the reasons why people act the way they do. Autonomy then focuses on the ability of persons to engage with moral ideas and to introduce a change in their moral repertoire subsequent to these engagements. In short, it accords credence to the capacity of persons for moral engagement and of preferences formulation.

Recognising the capacities of persons to be autonomous in oppressive circumstances recognises the inability of persons to act in accordance with their preferred preference. It concentrates on the process through which persons put into effect their preferred preference, and focuses on the ideas that lie behind action. Recognizing ideas that may not be expressed in action concentrates on the ethical reflection involved in the privileging of a particular preference and the role of external circumstances in determining the course of a particular action. It, therefore, makes a distinction between the agentic capacities that persons may develop and their ability to exercise these capacities, between ability to develop their agentic capacities and the actual conditions within which persons live. However, to make this distinction is not to say that the conditions within which persons live are unimportant. On the contrary, our capacities develop and are sharpened when we exercise them. While ideally persons should be able to live in conditions where they are able to develop and exercise their agentic capacities, the actual conditions within which many persons live, especially the circumstances of most women, are marked by subordination. The question that follows is if we reduce our reliance on action as the primary expression of beliefs and desires, then how do we access our 'preferred preferences'? Our preferences, I suggest, can be accessed within our speech practices. This works on the idea that upon investigating the reasons for our action, we can identify the reasons for undertaking a particular action and articulate a preference for an alternative course of action where the immediate circumstances differ. When speaking of ourselves, our goals and aspirations, we are able to reveal not only what our preferred option is but also the reasons for abandoning this preferred course of action. In examining speech practices, therefore, we are able to explore possibilities which our actions conceal and disguise.

This theoretical argument outlining the possibilities of recognising autonomy within oppressive contexts has an empirical basis. It is based on my fieldwork conducted among the women primary/village level workers known as the *Sathins*[5] who are involved in a state sponsored programme for women's development in the northwest Indian State of Rajasthan. The *Sathins* are largely illiterate or semi-literate and belong mainly to the lowest castes. The fieldwork was conducted in the districts of Jaipur and Ajmer over a period of eight months, between September 1998 and April 1999. The fieldwork examined the processes through which persons recognise their subordination and develop capacities and responses to challenge it. In short, it focused on the agentic capacities of persons within oppressive circumstances, on capacities, which were not always evident in their actions. Thus the women demonstrated desires, preferences on a wide range of subjects from issues of childbirth to political participation and recounted their experiences with respect of these. Their desires and preferred preferences in most cases were in remarkable contrast to their actions.

Let me illustrate what I mean by focusing on the ideas behind actions with the following examples. These illustrations reflect the common fertility-related conflicts that many of the women studied in the fieldwork experienced, between their preferences and their expected roles. To point this out is not to assume that women have an 'alternate set' of preferences which are in clear opposition to what is popularly expected of them. It is to simply state that there are often no 'neat preferences' that either easily translate into action or crystallise as our 'preferred preference'.

Example 1: Interview with Geeta

> I was the first woman in my village to get operated [sterilised] around fifteen years ago. I already had two boys and two girls and did not want anymore children. When the jeep (of the family welfare department) came, I got in and sat there all alone. Nobody accompanied me. My husband was not at home at that time as he had gone to another village. I went ahead with my decision without his knowledge. My mother-in-law never gave me anything to eat after I came back from the hospital. After ten days, when I came back after having my stitches removed my mother-in-law ordered me to get down to work. I went to the fields for the *lavani* [wheat harvesting].[6]

Example 2: Interview with Mohan Kanwar

> I have five daughters and one son. My mother-in-law did not allow me to undergo sterilisation. Once I had even climbed on to a tractor that was taking the village women to the sterilisation camp organised by the government health department. My mother-in-law came running after me, bringing with her a *lathi* [wooden stick] and beat me into climbing down. She was very strict and kept us under such strict regime. While my husband agrees with me, he dare not say a word to oppose my mother-in-law.[7]

Example 3: Interview with Sohan

> Our *Sathin* has spoken to us about a lot of things. About observing cleanliness, about sending our children to school, about not discriminating between the boys and the girls. We have understood but our husbands do

not understand. I have two children but I want to have more children as I do not have a son. I just have two daughters. My youngest daughter is five years old and I have been taking contraceptive pills in order not to get pregnant. My husband will not agree to stop having children till I have a son. I am a member of the women's saving group in the village and I deposit twenty rupees a month in the saving group. The money is earned by me, where else will I get the money from? My husband does not give me any money. I make ghee (clarified butter) from the cows and sell it and some of that money I deposit in the savings scheme.[8]

The first two examples are fairly straightforward and provide clear insights into the respondents 'preferred preference'. In the case of *Sathin* Geeta, she is able to translate her preferred desire into her actions. On the other hand, Mohan Kanwar, is physically prevented from achieving her preference. The case of Sohan is more complex. While she wants to have more children in the hope of having a son, she however takes precautions in order to avoid further pregnancies. The decision not to have any more children however, she makes clear, will finally depend on her husband and on her ability to produce a son. It is interesting to note here that while she is able to finance her contributions to the village women's savings group through her own economic efforts, she is unable to formulate a clear preference in respect of her fertility options.

The dilemma that these narratives pose for an exercise documenting female autonomy is whether these responses count as an evidence of the autonomous capacities of these women. The popular understanding of autonomy as 'doing what one wants' with an emphasis on control and power over one's life would be dismissive of at least two of these responses. However, the significance of these responses lie not in the abilities of these women to undertake free action but rather in their reflexive capacities to actively formulate and articulate their preferences. Any exercise documenting women's agency, particularly within oppressive contexts, must take these capacities into account.[9]

Part II

Demography and Women's Autonomy

How has demography treated the issue of women's autonomy? To an unspecialised reader, this may seem a confounding question. 'Why should autonomy be any concern of demographers?' they might ask. Autonomy is increasingly finding a prominent place within demographic research. The concept of autonomy, demographers believe, has an impact on demographic outcomes, and is generally purported to achieve the following demographic goals:

- postponing the marriageable age of women
- motivating women to limit fertility within marriage
- spreading the costs of fertility regulation more equally amongst men and women.[10]

It is assumed that an increase in women's autonomy would lead to a postponement in their age of marriage, it will increase women's economic independence which in turn will lead to a decline in son preference, and women's autonomy as manifested in their increased knowledge of modern contraceptive options would lead to fertility regulation, increased use of contraceptives and hence fertility decline. (Mason 1993, Morgan and Dharmalingam 1996).

In this section, I shall examine the conception of autonomy privileged by demographers. The demographic conception of autonomy constitutes what I shall call an 'autonomy worry'. While this paper isolates and focuses only on the manner in which autonomy is employed within demography, it is important to note that demographic research on women has not been immune to criticisms. It has been argued that biological and gendered assumptions underlie demographic research. These include: different expectations from men and women, the meagre range of activities classified as 'women's activities' or 'feminine activities', the exclusive focus on women in fertility studies led by the assumption that women are the sole producers and carers of children, the tendency to look at women's behaviour as representative of a separate women's culture, of applying linear theories of modernisation and development (Watkins 1993, Hayes 1994), of being ahistorical, acontextual and arbitrary. And finally, it is argued that reproductive behaviour and practices as 'extracted from their real context and fitted into abstract, metaphorical histories that all societies are said to undergo' (Greenhalgh 1995: 20).

The 'autonomy worry' within demographic literature has two main components: the first relates to the conceptualisation of autonomy within demographic literature and the second concerns the employment of quantitative techniques to measure/recognise autonomy. These two main components of the 'autonomy worry' unpack into five main problems affecting this literature. These are the following: the model of the autonomous self as a self-interest maximising individual is problematic and oblivious to feminist concerns. The model of autonomy employed in this literature is a substantive one. Fertility studies employ a substantive conception of autonomy in two ways. First it accords a priority to certain substantive ideals such as independence and consequently, autonomy accounts within fertility studies tend to look for decisions or choices which display substantive independence. Second, these studies place a certain value on the content of these choices. In other words, these choices have to be of a specific content in order to be classified as autonomous. For example, desiring lower fertility is regarded to be evidence of high autonomy and its opposite an indicator of low autonomy. In addition to adopting substantive ideas this scholarship does not distinguish between autonomy and freedom. It presents descriptions of cultures, which do not idealise autonomy, rather than an account of how persons fashion their response either to forces of socialisation or the oppressive features of their environment. The quantitative techniques that are used to 'measure' women's autonomy are deeply problematic. Apart from the intrinsic problems of quantifying such an attribute as autonomy, these techniques turn women and their capacities into an indistinguishable statistical mass. Finally, this literature is sociologically naïve and leaves little or no space for an account of women's agency in its analysis.

Two questions might be asked here. Why does autonomy appear as an 'independent variable' in demographic research and why is there a problem with quantitative techniques when demography is an empirical discipline? The use of the autonomy variable in explaining fertility behaviour is linked to and is a product of the shift in demographic research in the 1980s. Classic demographic transition theory, which influenced demographic research in the 1950s and the 1960s, came under close scrutiny with the publication of the Princeton project findings. This project commissioned in 1963 spanned two decades (and included John Knodel and Etienne Van de Walle, Susan Watkins, Barbara Anderson[11]) and put the propositions of the classic theory to test. Its findings that supposed indicators, urbanisation, literacy, infant and child mortality and industrialisation failed to account for the pattern of decline in the various regions of Europe heralded a 'new era'[12] in demographic research. This new era, marked by the shattered faith in the transition theory,[13] ushered in interest in the new demographic variable, *culture*. 'Cultural setting' it was noted, 'influenced the onset and spread of fertility decline independently of socio-economic conditions' (Knodel and Van De Walle, 1986). Culture as a demographic variable 'operationalisable' within language, ethnicity and geographical location led to the cultural or diffusion approach to fertility decline. This era in demography, which continues till today, is marked by 'a self conscious search for methodologies that will allow demographers to incorporate cultural meanings into their explanations of demographic processes' (Fricke 1997: 825). It was to incorporate culture as a demographic variable that led demographers to turn towards anthropology. The preoccupation with cultural meanings led to yet another development, the focus on women's roles, their status and lately their autonomy as an explanatory variable in the understanding of fertility decline. Autonomy, a complex concept internally differentiated and difficult to characterise and indeed define, is employed within this literature not only as a variable with a consistent meaning but one that is also amenable to quantification. There is a link between the two. The meanings that demographers attribute to autonomy influence their designs of the quantifiable 'indicators' of autonomy.

The quantification of autonomy is deeply problematic. There are two problems associated with using quantitative techniques for the purpose: Firstly, autonomy of persons cannot be captured or indeed measured through demographic techniques such as correlation analysis. The crux of the problem is whether autonomy of persons can be judged by focussing on one or more activities at a fixed time in their lives. For instance can we enter judgments on autonomy of persons on the basis of their fertility activity?[14] While persons can be regarded as having developed certain autonomy enhancing qualities, which may include amongst others, a commitment to a set of values, motivations, decision-making capacities, reflexive capacities, etc., it is unclear within demographic scholarship which amongst these are admitted or ignored in judging the autonomy of persons. Secondly, there are thorny questions of responsibility and what constitutes responsible and 'conformist choice', requiring an attention to issues of procedural independence, all aspects of moral autonomy of agents which are not likely to be resolved by statistical interventions. Procedural independence refers to an analysis of the procedures through which certain decisions are made, placing an emphasis on the independence of the process through which these decisions are made. The statistical data collected fails to make a distinction between the exercise of 'genuine' choices and those that are made on the behest of others. For example, on the question of contraceptive use, how can we be sure on the basis of statistical evidence that the choice of contraceptive or even of the decision to use contraceptive methods is made by the woman herself? This example also does not also reveal what the 'real or actual' desires[15] of the woman in question might be. Perhaps, she might want to have more children and that the decision to limit her fertility was not what she desired in the first place. Can demographers attribute purposes to these women's actions and say that high fertility or the desire to have more children is what they truly want.

In addition to serious questions about the quantification of autonomy, it might be argued that judgements on autonomy, a characterisation of personhood and of moral capacities cannot be attempted on the basis of decisions covering a very narrow albeit important area of ones life, i.e., fertility (despite the attempts of demographers to accord significance to it). Demographers side-step this objection by stating that autonomy is influential in determining fertility behaviour. This fertility determining autonomy is quantified as the absence of external constraints within a social context. However, these operationalisable accounts of autonomy are not referring to autonomy at all. At best what is invoked here is the idea of negative freedom or the freedom from external constraint.[16] Negative freedom or liberty claims that the freedom of a person is determined by the restraints that a person encounters in the carrying out of her decisions. It does not pay attention to the person and her complex set of capacities of formulating her desires, values and choices, etc., in the calculation of her freedom.[17] Free agents according to the proponents of positive liberty must be in some senses self-governing and their external autonomy consist of more than freedom from external constraint.

In order to operationalise[18] autonomy as a variable in demographic research, demographers construct certain indicators in an effort to capture the aspects of autonomy which may reflect and impact on fertility behaviour but also which can be easily quantifiable. These indicators are used by demographers to make evaluative judgments on the level of autonomy enjoyed by women in these societies. Finally, these indicators are important resources shedding light on the understanding of autonomy employed by demographers. Autonomous persons are understood in this literature as excessively individualistic, rationalist, separate and decontextualised. This model of the self, which is reflected in the indicators, is oblivious to feminist concerns about this model of the autonomous self but also to the recent theorising on autonomy that does not insist on substantive independence.

Let me give a few examples of this employment of negative freedom as autonomy within demographic literature. This I shall do so by examining what are known as the 'indicators of autonomy'. These indicators are quantitative in nature and are used by demographers to make evaluative judgements on the level of autonomy enjoyed by women in these societies.

Autonomy as an Instrument, Autonomy as Negative Freedom: Examining the 'Indicators' of Autonomy

The indicators of autonomy developed by demographers emerge from their understanding of autonomy. Autonomy understood in terms of negative freedom informs these indicators, and this can be ascertained by their definitions of autonomy. This literature then alludes to autonomy in several ways. Women's position, their power, independence and control are all used interchangeably with autonomy.[19] A number of demographic hypotheses are based on women's autonomy. These are mainly related to child mortality, reduced fertility etc. For example, 'Women's autonomy and economic independence contribute to child survival by increasing the mother's ability to provide her children with adequate nutrition and child care' (Mason 1993: 25)

The most used definition of autonomy within the demography of women is the one developed by Dyson and Moore (1983). According to them, female autonomy is the 'capacity to manipulate one's personal environment.

> Autonomy indicates the ability – technical, social and psychological – to obtain information and to use it as the basis for making decisions about one's private concerns and those of one's intimates ... Societies in which females have high personal autonomy relative to males are typically characterised by several of the following features; freedom of movement and association of adolescent an adult females; post marital residence patterns and behavioural norms that do not rupture or severely constrain social intercourse between the bride and her natal kin; the ability of females to inherit or otherwise acquire, retain, and dispose of property; and some independent control by females of their own sexuality for example, in the form of choice of marriage partners. (Dyson and Moore 1983: 45)

Jeffery, Jeffery and Lyon (1988) build upon Dyson and Moore's framework but they look specifically at the indicator of autonomy that relates to women's close natal links. They posit then that women who have close natal ties, evident in their visits to their mother's home after marriage, enjoy greater freedom of movement and autonomy. Many other demographic writings have come to develop similar indicators of women's autonomy. Common to all of these indicators is the emphasis on women's freedom of movement and an important role in family decision making. A typical autonomy argument is: 'greater freedom of movement and a more important role in family decision making imply greater autonomy'[20] (Niraula and Morgan 1996).

Leela Visaria (1996) attempts to evolve 'measurable female autonomy indices' (1996: 268) and identifies the following indicators of autonomy: income autonomy, women's ability to perform certain tasks on their own – personal autonomy, contact with natal kin and level of education. For example, personal autonomy included women's independent decision-making ability in matters of buying clothes or other personal items for the self or for children, taking a sick child to the doctor, etc. If the woman did not seek the permission of the mother-in-law or the husband in doing any of these she was regarded as being autonomous and given a score of one. Conversely, if a woman sought permission in the performance of any of these tasks she was regarded to be non-autonomous and a score of zero was marked against her. (Notice that in this example, there is no mention of the distance of markets from the home or if there were any modes of public or private transport available to these women). Vlassoff (1996) employs four indicators to measure autonomy, (a) exposure to the outside world, (b) decision making regarding choice of marriage partner, dowry and in domestic matters. In the same volume, Irudaya Rajan et al. (1996) plot the following 'objective' indicators on their autonomy scale. These are: proportion of women owning property, proportion of women having their own source of income, proportion of women receiving the income they earn and retain their income, proportion who retain cash independent of their husbands, percentage of women who have their own savings account and proportion of women who can decide on their own regarding, (a) daily household purchases, (b) purchase of personal goods like saris or chappals, (c) purchase of children's clothes, (d) taking a sick child to the doctor. According to Jeejebhoy (1995), apart from women's ability to obtain information in order to make decisions about oneself and one's intimates, autonomy also includes the active role of women in family and in society, their decision-making and execution powers, whether they are free to develop bonds with their husbands, freedom of movement and interaction with the outside world. In the context of West Africa, Etienne and Nicholas Van de Walle (1993) describe autonomy as beneficial to demographic outcomes. According to them:

> Autonomy benefits the mother and her children because it renders her capable of taking decisions alone. An autonomous young woman will decide by herself to look for help when her child is sick, she will treat her daughters the same way as she treats her sons, she will take decisions at least in her feminine world. Autonomy has to do with control over resources even when it is vicarious control, by the goodwill of a male household head, in the sphere of activity where he has delegated authority.
> (Van de Walle 1993: 63)

They detail women's autonomy in respect to the control that women exercise over the natural determinants of fertility. These natural determinants of fertility include what are known as the 'high- priority proximate determinants of fertility',[21] which include exposure to the risks of pregnancy, entry into conjugal union, and polygyny and what they refer to as the ABC complex, which includes abstinence, breast feeding and contraception. According to them, while recourse to modern social practices such as monogamy does not necessarily result in lowered fertility, however, 'the question of women's autonomy in matters of reproduction must be answered in detail by looking at the various proximate determinants of natural fertility. On all counts, it appears that women do little more than inhabit, and to some extent administer, a domain which belongs to men' (Van de Walle and Van de Walle 1993: 79).

It is evident from the above indicators that this literature concentrates on the aspects of women's lives that are considered demographically important and draws conclusions on their autonomy from these.

The 'indicators' of women's autonomy are measured through employing several kinds of field techniques, quantitative and qualitative. The field questionnaire appears as a popular choice to measure autonomy. These questionnaires are carefully designed and closely controlled to smooth over the complexities in their respondents choices, producing in turn, a neat set of responses. However, we are complex beings and do not only think in terms of binary opposites or either/or choices. We choose from across categories and do not always endorse wholly one set of goods and discard all of the goods contained in another set.

Another measure of women's autonomy concentrates on her levels of literacy.[22] In addition to age and marital status, literacy is the most employed variable in the fertility analyses in the developing world.[23] The importance of literacy for women's autonomy is invoked by demographers in order to establish the link between increased autonomy of women as measured through their literacy and its impact on lowered fertility. It is assumed that female literacy would lead women to be informed about not only the ways of preventing repetitive and several pregnancies but also make them aware of the benefits of small family sizes.[24] It is not hard to establish a relationship between autonomy and literacy. Similar assumptions underlie both. Literacy like autonomy is associated with ideas of 'progress', 'civilisation', 'individual liberty', 'abstract thought', 'logic', 'scientific thought' and 'social mobility', 'modernisation syndrome', and the concept of the 'modern man' the 'development of empathy', 'flexibility', 'adaptability' and 'willingness to accept change', etc.,[25] (Oxenham 1980, Street 1984).[26] There are quantitative techniques adopted to measure levels of autonomy among literate and non-literate women and analysis put forward correlating women's fertility levels or aspects of their autonomy to their literate or non-literate status. Educating women results in what have been termed as 'direct and indirect benefits' (Chanana 1996: 110). The direct benefits of education include higher marriageable ages of and increased employment of women (outside the home). The indirect benefits of education include the development of new or 'modern' values and attitudes (Cleland and Wilson 1987, Sathar et al. 1988, Weller 1984). Educated women, it is generally believed, are more inclined to value their daughters, adopt a small family size, exhibit a reduced dependence on their sons and thus the accepted correlation between high educational attainment and high autonomy (Chanana 1996: 10). Education not only facilitates these transitions but the higher the educational attainment; the higher will be the degree of autonomy.'

There are problems with assuming that literacy by itself would lead to women's autonomy. There is no discussion on the specific kinds of literacy skills and practices persons must acquire in order for them to increase their autonomy. Schools encourage uniformity and acquiescence. They reinforce prevalent gendered roles and they are therefore far from instilling qualities of self-determination.[27] Furthermore, the schooling of girls is also a socially sanctioned good and therefore, access to schools is subject to existing social agreements on this respect. These discussions revolve around the instrumentalist connection between literacy and autonomy and do not examine the relationship between the subjects and their literacy skills, of how they perceive their own 'literate status'. Finally, many of these arguments establishing a correlation between literacy and fertility decline are based on the assumptions which characterise the illiterate woman as lacking in both the capacities and knowledge of what constitutes the well being of their children. Illiterate women are characterised as solely involved in the activity of reproduction, their education it is claimed would provide them with opportunities to pursue other activities! This not only misrepresents the multiple tasks that women are involved in both in the household, and in the income generating activities, but also devalues their parenting roles.

Underpinning this direct correlation between low autonomy of women and the presence of certain social practices and structures is the assumption that were these constraints removed persons would behave and act differently. This theorising then does not take into account three possibilities:

1) The possibility of persons (women in this case) of not committing the action (of lowering their fertility) despite there being less restrictions on their choice and the presence of alternatives
2) The possibility of committing certain actions despite the threat of violence, coercion, etc.
3) The possibility of people willing themselves to remain 'inactive' and submitting themselves to the will of others and being able to offer reasons for doing so.

Demographic studies of women's fertility view autonomy primarily as the absence of external constraint to free action. Consequently, autonomy is privileged as a value characterised by free action, or the ability to transform one's desires into action in an unobstructed environment. Thus, the main question in demographic literature is wrongly posed. It seeks to ask questions relating to the societies in which autonomy of women is higher or absent. Depending upon the levels of social and personal freedoms considered legitimate in every society, it arrives at an analysis derived from a simple correlation model. This line of questioning corresponds to issues relating not to when persons are autonomous but rather to the conditions in which persons are free to exercise their autonomy. Therefore, the discussion is centred around enumerating the possible restrictions on the exercise of choice rather than on the formulation of choice and its social and moral foundations.

Instead of presuming that agents would act differently if the external constraints were removed, the question should be: In what ways would the person exercise her capacities given the removal of external constraints? Would persons act in radically different ways if those external constraints were removed? To presume that women would desire fewer children, or even girls, were the constraints on women's free movement or education lifted would be to assume two things:

1) That women are separate from the society that they live in and that they somehow have a separate morality
2) The second would be to assume that somehow women have no stake or a limited stake in the perpetuation of existing morality.

This literature relies on freedom of action and its accompanying model of the self as the autonomous 'chooser' who is responsible for her actions. This reliance on action is not concerned with the degree to which actions of persons reflect their moral choice or are a response to coercive circumstances or committed under the influence of others. Further, action marks the end of our journey in terms of thinking about values and judgements. Action does not form the core or even perhaps our preferred judgement or motivation or idea. The emphasis on action as the governing measure of autonomy then fails to capture the 'actual or real' desires that might lie behind action, desires that might be different from those which motivate action. Therefore there is a need to determine the ideas that lie behind action.

There is some ambiguity in the literature where the source of responsibility lies for the prevalence of high fertility. Does it lie at the level of the social or does the burden of displaying autonomy fall on the individual? There is a tension between the excessively over-socialised, passive view of the person that is presented in this literature and the demands of autonomy that they are supposed to fulfil. Despite this prevailing tension, the responsibility for high fertility tends to be placed upon the passive woman, as it is her autonomy that is the focus of study. This judgement on her supposed heteronomy, I argue, is more a judgement on the low rank of freedom (in comparison to the other goods) in society than on her capacity for autonomy. While some amount of social freedom is required to 'exercise' one's autonomy and indeed to develop it; however, it is not always the precondition for possessing the capacity for it. Premising female autonomy upon the social agreement on the kinds of freedom where women are almost never contributors, demographic literature does not leave any space for women's agency. Moreover, this literature presents a static view of the self (women). They are portrayed as 'passive victims of patriarchal institutions who have little choice but to surround themselves with children' (Greenhalgh 1995: 25). By rendering them non-autonomous on the basis of their actions, this literature denies women their capacities for autonomy. In denying this capacity they are oblivious to their ability to bring about change in their lives as well as their capacity for reflection and of imagination – of imagining an alternate life and roles for themselves.

'Demography of Women' within Conditions of Subordination

It was proposed in Part I of the paper that the agentic capacities of persons can be recognised even within conditions of subordination. The recognition that persons possess agentic capacities (even in the absence of certain conditions) has implications for the academic projects characterising vast sections of the population, particularly women as passive beings, but also on the nature of the policy discourse. A respect for the capacities of persons must be evident in the creation of conditions within which persons can exercise their capacities. A respect for individual autonomy within fertility policies for example, must be reflected in a concern for providing for physical and monetary needs as well as the information needs of persons in order for them to exercise their autonomy. A respect for autonomy within policy would rest on the premise that the best way for building responses to oppressive circumstances is to change the conditions themselves and not expect individuals to adapt to them in some private way.

Looking for ideas behind action then would lead us to develop qualitative indicators of autonomy. These qualitative indicators would seek to see for example, how a person's preferred choices are in contravention to the choices which they put into practice. A qualitative analysis of women's childbearing decisions, for example, would focus on who decides, when and how many children to have, but also ascertain the level of identification (of the women in the study) with each of these decisions. It would attend to the presence of an alternative set of preferences and identify the factors such as familial, social and institutional, which obstruct the success of these preferences. More importantly it would take into account the self descriptions and self understandings of the persons in question. Let me explain what I mean by these self-descriptive accounts from an example from my fieldwork amongst rural women belonging to two districts in Rajasthan. Many of the women I interviewed described themselves as *padhi-likhi* or literate.[28] In fact, they were at best semi-literate. They were referring however, to the particular kinds of knowledge or *Jaankari* possessed by them, which consisted of among other things, knowledge of state institutions, hierarchies and procedures. I have referred to this particular capacity elsewhere as 'political literacy'.[29]

Conclusion

It has been suggested that a principal problem in studying the impact of women's autonomy on demographic outcomes in developing countries lies in the non-availability of the word autonomy itself within the vernacular. This is said to be particularly true of South Asia.[30] While this is an excessive claim, it is not without some merit.[31] However, the problem is not only one of non-availability of corresponding literal terms within different languages but also of different concepts, meanings and values across different moral and cultural contexts. In relation to the autonomy of persons within fertility studies, this problem is magnified in part mainly because of the manner in which autonomy has come to be understood in this literature. Autonomy, understood as a character trait opposed to interdependence, as symbolic of a rational self-sufficient, self-realising individual with clearly set plans and goals (low fertility in this case) can be elusive in itself, particularly so in social contexts where there exist a number of constraints on women's freedom. How do we resolve this problem of non-translatability of concepts? One way is to embark on a qualitative exercise of looking for abilities, capacities that would be resonant in both the settings. It is then possible to determine autonomy in terms of capacities of persons to form second order preferences about their desires. Their capacity for self-reflection as well as to institute a process of questioning of their moral values in the light of perhaps, some new ideas encountered by them. Their capacity to effect change, change that may not always be evident in action but through their speech practices. In other words, looking at autonomy means not merely looking at actions but looking instead at the ideas, which propel certain kinds of action or non-action.

Notes

1. I am grateful to Dr Sudipta Kaviraj for his detailed comments on earlier versions of this paper. I would also like to thank Maya Unnithan-Kumar for comments and editorial suggestions towards improving the clarity of the paper. The discussions on an earlier draft of the paper at the workshop on 'Anthropology and Reproductive rights' University of Sussex were most useful. I thank all the participants and audience for their comments. The fieldwork was made possible by a doctoral fieldwork grant made available by the 'Central Research Fund', University of London and the Inlaks Foundation.
2. It must be pointed out that referring to the subordination and inequity in the majority of women's lives does not imply that these conditions have somehow been historically static or that these societies continue to be socially stagnant. On the contrary, the emphasis on autonomy enhancing

capacities within subordinate circumstances is a recognition of the existing possibilities of questioning and challenging the apparent and very real practices of oppression.
3. While, there is no singular definition of autonomy, the lack of consensus surrounding its definition has led to a web of meanings woven around the idea. At a broad level however, it invokes capacities as well as achievements. Capacities to formulate preferences and an ability to translate into action or achieve our choices, which our capacities help us formulate. Thus far there seems little within this characterisation to dispute. It is when the capacities are infused with a particular substantive content, i.e., when the outcomes of these capacities are specified that disputes arise. A popular dispute for instance, concerns the characterisation of autonomy as an ideal that facilitates independence; the disengaged separate self unmoved by personal or social circumstance.
4. See Marilyn Friedman (2000) and Gerald Dworkin (1988).
5. The word *Sathin* literally translates as 'female companion'.
6. Interview with Geeta, *Sathin*, Village Palu, District Jaipur, 14 December 1998.
7. Interview with Mohan Kanwar, *Sathin*, Village Mandaliya, District Jaipur, 20 March 1999.
8. Interview with Sohan, Tootoli village, District Jaipur, 20 December, 1988.
9. In a recent study of women's reproductive experiences, Maya Unnithan-Kumar (2001) analyses the role of women's 'emotional attachments' in influencing the reproductive health care decisions of rural Rajasthani women. In a study of the motivations underlying their decisions, she suggests that although there is a 'tendency of the women in (rural Jaipur) towards inaction in seeking health care services', this could not always be seen as symptomatic of a lack of desire. 'Yet, given a certain level of support, particularly from kin, they were also willing to seize the opportunity of exploring further possibilities of health care' (2001: 35).
10. See Mason 'The Impact of Women's Position on Demographic Change During the Course of Development' (1993). Mason expands these ideas on the impact of women's position on the fertility transition as:

    Effects by way of women's age at marriage
    1) An increase in women's autonomy will facilitate the postponement of marriage and hence the decline of fertility by reducing the need to control unmarried women's sexuality through early marriage.
    2) In family systems that give all rights of women's labour to the husband's family, women's economic independence will facilitate the postponement of marriage and hence the decline of fertility; in other family systems, the effects of women's economic independence on marriage are indeterminate.

    Effects by way of motivation to limit fertility within marriage:
    3) Because it channels the rewards of children disproportionately to men and the costs of realign them disproportionately to women, patriarchal family structure encourages high fertility; egalitarian family structure facilitates fertility decline.
    4) Women's economic dependency on men produces strong son preference among both women and men and hence relatively high fertility desires for purposes of risk insurance and old age security; in a conjugally

oriented family system, women's economic independence facilitates fertility decline.
5) The extent of which women's autonomy and economic dependency determine women's dependency on the maternal role for legitimacy, security and satisfaction, and hence the opportunity costs of having children and the motivation to limit fertility.

Effects by way of costs regulation:
6) women's autonomy influences their access to modern knowledge and modes of action and hence their propensity to engage in innovative behaviour, including fertility limitation within marriage.

The first two hypothesis concern the impact of women's autonomy and economic independence on their age of marriage, itself a proximate determinant of the fertility transition in contemporary developing countries. The first idea is that women's age at marriage is directly linked to their autonomy, because early marriage is a strategy used by family elders to control the sexuality of females. Thus, an increase in women's autonomy is likely to mean that pressures for early marriage of women have weakened (Mason 1993: 30-31).

11. A. Coale and Susan Cotts Watkins, eds. *The Decline of Fertility in Europe*. Princeton; Princeton University Press, 1986.
12. See Tom Fricke (1997)
13. The other criticisms of transition theory are: '1) its overstylised and incomplete account of the major determinants of demographic change 2) an inaccurate depiction of the historical process, the actual pathway of demographic change 3) the inattention to the group; especially the class specificity of demographic change; and 4) ethnocentric assumptions about the units of demographic decision making and behaviour.' See Greenhalgh, 'A Political Economy of fertility'. *Population and Development Review*, 16, No. 1 (March 1990).
14. As mentioned in Part I, philosophers are not agreed on a 'definition' of autonomy, preferring instead to 'characterise' the complex concept. Autonomy most autonomists agree, is one of degree and persons can only be 'more or less autonomous' (Raz 1986: 154).
15. By 'real and 'actual' choices I am simply referring to the choices that persons may hold or make in the event of there being different circumstances or those which persons may make despite different circumstances.
16. I owe the clarification of this point to Nick Hostettler (personal communication). The invoking of negative freedom in order to make claims about autonomy is deeply problematic. The distinction between negative and positive and negative freedom has been a subject of controversy and debate, not least between Isaiah Berlin and C. B Macpherson. Other philosophers have distinguished between freedom and autonomy, most notable among them being Gerald Dworkin. Theorists who have opposed this distinction between positive and negative liberty include amongst others Gerald Macculum (1976).
17. According to John Christman (1991) 'the purveyors of the notion of positive liberty insist that the person and her capacity to formulate her desires, values and goals is a crucial element in the calculation of her freedom' (Christman 1991: 343). See also Charles Taylor (1979).

18. See for instance Jeejebhoy (2000). Jeejebhoy describes the aim of her chapter as one that seeks to 'operationalise what is meant by women's autonomy' (2000:204).
19. For example, Alaka M. Basu (1996) states autonomy in its simplest form to be 'the freedom or ability to make decisions on a given matter, or the right to exercise of choice' (Basu 1996: 52). According to T. K. Sundari Ravindran (1999) autonomy means 'control over significant decisions affecting their (women's) lives, and having access to resources that would enable them to do so' (Ravindran 1999: 35). Consider the definition of power offered by Constantina Safilios-Rothschild (1982). She writes 'Female power can be defined as the women's ability to control or change other women's and men's behaviours and the ability to determine important events in their lives, even when men and older women are opposed to them' (1982: 117). On the conflation between autonomy and authority see, Kritz and Makinwa Adebusoye (2001). On using autonomy and empowerment interchangeably see Jeejebhoy (2000) according to whom, 'autonomy and the more commonly used word, empowerment attempt to capture similar dimensions of women's situation and are often used simultaneously. Definitions of women's empowerment and autonomy appear thus to converge as far as the end is concerned; gaining control over their lives vis a vis community, society and markets' (Jeejebhoy 2000: 205).
20. Similar measures of autonomy involved are: (a) perceived economic independence, i.e., whether a woman reports that she can support herself and her dependents without her husbands help, (b) freedom to move within and between villages, (c) spousal interaction, i.e., whether the spouses discuss family finances and desired family size. For instance see Morgan and Dharamlingam (1996).
21. J. Bongaarts (1985) 'What can Future Surveys Tell Us about the proximate determinants of Fertility?'
22. Jeejebhoy (1995) for example, identifies five linkages between women's fertility and her literacy. These are knowledge autonomy, decision-making autonomy, physical autonomy, emotional autonomy and Economic and social autonomy and self-reliance.
23. Caroline Bledsoe, John B. Casterline, Jennifer A. Johnson-Kuhn and John Haaga, eds (1999).
24. See for instance, Mamtha Murthi and Jean Dreze (1999). The authors contend that in respect to India, women's education was the most important factor in explaining fertility differences across the country. Female education can be expected to reduce fertility behaviour for the following reasons: 1) education raises the opportunity cost of women's time and opens up greater opportunities for women that often conflict with repeated child bearing. Education may reduce son preference directly bearing on reduced fertility as large families are often a desire for higher incidence of son survival rates, educated women have higher aspirations for their children, educated may be receptive to modern social norms and family planning campaigns, female education reduces infant and child mortality and female education may result in achieving the planner number of births by assisting in the dissemination of knowledge and access to contraceptives (Murthi and Dreze 1999: 5).
25. According to Street (1984), the traditional representation of different cultures as logical and prelogical/primitive/modern and concrete/scientific

cultures gets replaces (within this literature) by that between literate and pre literate societies.
26. For detailed accounts of the connection between literacy and women's autonomy see Basu and Jeffrey (1996). For arguments drawing direct connections between literacy and fertility outcomes see John C. Caldwell (1980) H. J. Graff (1979), B. L. Wolfe (1980) and John D. Kasarda, John O. G. Billy, and Kirsten West (1986). According to Kasarda et al., 'First and foremost, schooling increases a woman's knowledge and competence in virtually all sectors of contemporary life. It broadens her access to information via the mass media and printed material. It develops her intellectual capacities and exposes her to interpersonal competition and achievement. It gives her an opportunity to acquire marketable skills and other personal resources to pursue non familial roles. It raises her image of her potentials and those of her children, and it simultaneously imparts her with a sense of efficacy and trust in modern science' (Kasarda et al. 1986: 88). Also quoted in Chanana (1986).
27. See Sharon Bishop Hill in Richard Wasserstrom, ed. *Today's moral Problems'*. Macmillan, 1979. Brian Street writes, that the examples available on the operation of literacy in different societies show it to be more often than not 'restrictive and hegemonic and concerned with instilling discipline and exercising control' (Street 1984: 4).
28. Interview with Mohan Kanwar, Village Mandaliya, district Jaipur 20 March 1999.
29. See my 'Autonomy, political rights and the social woman: towards a politics of inclusion?', in Subho Basu and Crispin Bates (eds). *Rethinking Indian Political Institutions*. Anthem Press, London (forthcoming).
30. According to Alaka Basu and Roger Jeffrey (1996) 'The absence of a word to talk about autonomy makes it hard to decide whether to rely solely on women's own perspectives. If autonomy is the ability to act as one wishes, how do we make sense of findings which show that some women have very limited wishes they are able to achieve' (Basu and Jeffrey 1996: 25–26).
31. However, this assertion made by Basu and Jeffrey (1996) is not strictly true but it does point to significant issue. In their claim, it is unclear, for instance, which South Asian language she is referring to. There are several words that translate as autonomy within Hindi, for instance. These are: *swarajya, svatva, adhikara, svaya and shasana*. However, these translate mostly as political autonomy and are often employed within the Hindi language in an official sense, for instance to refer to autonomous councils or local government bodies. These are rarely used popularly in everyday language to refer to individual autonomy.

References

Basu, Alaka M. 1996. 'Girls Schooling, autonomy and fertility change: What do these words mean in South Asia', in Basu and Jeffrey, eds. *Girls Schooling, Women's Autonomy and Fertility Change in South Asia*. New Delhi: Sage, 48–71

Basu, Alaka M. and Roger Jeffrey 1996. 'Schooling as Contraception', in Alaka Basu and Roger Jeffrey, eds. *Girls Schooling, Women's Autonomy and Fertility Change in South Asia*. New Delhi: Sage, 15–47

Benhabib, S. 1995. 'Feminism and Postmodernism', in Seyla Benhabib et al., eds. *Feminist Contentions*. New York: Routledge

Berlin, Isaiah. 1969. 'Two Concepts Of Liberty', in his *Four Essays On Liberty*. Oxford: Oxford University Press

Bledsoe, C., Casterline, J. B., Johnson-Kuhn J. A. and Haaga, J., eds. 1999. *Critical perspectives on Schooling and fertility in the developing world*. National Academy press

Bongaarts, J. 'What can Future Surveys Tell Us about the Proximate Determinants of Fertility?' *International Family Planning Perspectives*, 11, 86-90

Bhuiya, A. and Streatfield, K. 1991. 'Mothers Education and Survival of Female Children in a rural area of Bangladesh'. *Population Studies*, 45, 253-64

Caldwell, J. C. 1980. 'Mass education as a determinant of the timing of fertility decline'. *Population and Development Review*, 6: 3, 225-55

Card, C. 1996. *The Unnatural Lottery*. Philadelphia Pennsylvania: Temple University Press

Chanana, K. 'Educational Attainment, Status Production and Women's Auotnomy: A study of two generations of Punjabi women in New Delhi' in Basu and Jeffrey, eds. *Girls schooling and Women's Autonomy and Fertility in South Asia*. New Delhi: Sage, 107-32

Christman, J. 1991. 'Liberalism and Individual positive freedom'. *Ethics*, 101, 343-59

Coale, A. and Watkins, S. C., eds. 1986. *The Decline Of Fertility in Europe*. Princeton: Princeton University Press

Code, L. 1987. 'Second Persons' in Marsha Hanen and Kai Nelson, eds. *Science, Morality and Feminist Theory. Canadian Journal of Philosophy*, Supplementary volume 13, 357-82

Dworkin, G. 1988. *The Theory And Practice Of Autonomy*. Cambridge: Cambridge University Press

Dharamlingam, A. and Morgan, S. P. 1996 'Women's Work, Autonomy, and Birth Control: Evidence From Two South India Villages'. *Population Studies*, 50: 2, 187-201

Dyson, T. and Moore, M. 1983. 'On Kinship Structure, Female Autonomy, and Demographic Behaviour in India'. *Population and Development Review*, 9: 1, 35-60

Fricke, T. 1997. 'The uses of culture in demographic research: A continuing place for community studies'. *Population and Development Review*, 23: 4, 825-32

Friedman, M. 2000. 'Autonomy, Social Disruption, and Women', in N. Stolar and C. Mackenzie, eds. *Relational Autonomy*. Oxford: Oxford University Press

Graff, H. J. 1979. 'Literacy, Education and Fertility, Past and Present: A critical Review'. *Population and Development Review*, 5: 2, 105-40

Greenhalgh, S. 1995. 'Anthropology Theorizes Reproduction: Integrating Practice, Political Economic and Feminist Perspectives', in S. Greenhalgh, ed. *Situating Fertility: Anthropology and Demographic Enquiry*. Cambridge: Cambridge University Press: 3-28

Greenhalgh, S. 1990. 'A Political Economy Of Fertility'. *Population and Development Review*, 16: 1, 85-106

Hayes, A. 1994. *The Role of Culture in Demographic Analysis: A Preliminary Investigation*. Working Papers in Demography, Research School of Social Sciences, Australian National University, Canberra

Hill, S. B. 1979. 'Self Determination and Autonomy', in R. Wasserstrom, ed. *Today's Moral Problems*. London: Macmillan

Jaggar, A. 1983. *Feminist Politics and Human Nature*. Totowa, N.J.: Rowman and Allanheld

Jeffrey, P., Jeffrey, R. and Lyon, A. 1988. 'When did you last see your mother?', in John C. Caldwell et al., eds. *Advances in Micro Demography*. London: Kegan Paul, 321-33

Jeejebhoy, S. 2000. 'Women's autonomy in Rural India: Its dimensions, determinants and the Influence of context in Women's empowerment, and demographic Processes', in H. B. Presser and G. Sen, eds. *Moving beyond Cairo*. Oxford: Oxford University Press, 204-38

Jejeebhoy, S. 1995. *Women's Education, Autonomy and reproductive Behaviour: Experience from Developing Countries*. Oxford: Clarendon Press

Kasarda, J. D., Billy, J. O. G. and West, K., eds. 1986. *Status Enhancement and fertility: Representative Responses to Social Mobility and Educational Opportunity*. Orlando: Academic Press

Kertzer, D. I. 1997. 'Qualitative and quantitative approaches to historical demography'. *Population and Development Review*, 23: 4, 839-46

Kritz, M. and Adebusoye, M. P. 2001. *A Couple Agreement on Wife's Autonomy and Reproductive Dynamics in Nigeria*. IUSSP, XXIV General Population Conference, Salvador, Brazil, 18-24

Macculum, G. 1976. 'Negative and positive freedom'. *Philosophical Review*, 76, 313-34

Mackenzie, C. and Stoljar, N. 2000. *Relational Autonomy: Feminist Perspectives on Autonomy, Agency and the Self*. Oxford: Oxford University Press

Macculum, G. 1976. 'Negative and positive freedom'. *Philosophical Review*, 76, 313-34.

Macpherson, C. B. 1973. *Democratic Theory: Essays in Retrieval*. Oxford: Clarendon Press.

Martin, L. G. 1987. 'Female education and fertility in Bangladesh'. *Asian and Pacific Population Forum*, 1: 3, 1-7

Mason, K. O. 1993. 'The Impact of Women's Position on Demographic Change During the Course of Development', in K. Mason, N. Federici and S. Sogner, eds. *Women's Position and Demographic Change*. Oxford: Clarendon Press

Meyer, P. 1991. 'Mother's hygienic awareness, behaviour, and knowledge of major childhood diseases in Matlab Bangladesh', in J. Caldwell, S. Findley, P. Caldwell, G. Santow, W. Cosford, J. Braid and D. Broars-Freeman, eds. *What we Know about health transition: The cultural, Social and behavioural Determinants of Health*. Canberra Health Transition Centre: Australian National University

Meyers, D. T. 2000. 'Intersectional Identity and the Authentic Self', in Natalie Stoljar and Catriona Mackenzie, eds. *Relational Autonomy'*. Oxford: Oxford University Press

Meyers, D. T. 1989. *Self Society and Personal Choice*. Columbia University Press, New York

Moore, M. and Dyson, T. 1983. 'On Kinship Structure, Female Autonomy, and Demographic Behaviour in India'. *Population and Development Review*, 9: 1, 35-60

Murthi, M. and Dreze, J. 1999. *Fertility, Education, Development: Further Evidence From India*. Centre For History and Economics, Kings College Cambridge

Nedelsky, J. 1989. 'Reconceiving Autonomy'. *Yale Journal of Law & feminism*, 7, 7-36

Niraula, B. B. and Morgan, S. P. 1996. 'Marriage Formation, Post-Marital Contact with Natal Kin And Autonomy Of Women: Evidence from Two Nepali Settings'. *Population Studies*, 50: 1, 35-50

Obermeyer, C. M. 1997. 'Qualitative Method: A key to a better understanding of demographic behaviour?'. *Population and development Review*, 23: 4, 813–18

Oxenham, J. 1980. *Literacy, Writing, Reading and Social Organisation*. London: Routledge, Kegan Paul

Potter, J. E. and Volpp, L. P. 1993. 'Sex Differentials in Adult Mortality in Less developing Countries: the evidence and its explanation', in K. Mason et al., eds. *Women's position and demographic change*. Oxford: Clarendon Press

Rajan, I. S., Ramanathan, M. and Mishra, U. S. 1996. 'Female Autonomy and Reproductive Behaviour in Kerala: New Evidence from the Recent Kerala Survey', in A. Basu and R. Jeffrey, eds. *Girls Schooling, Women's Autonomy and Fertility Change in South Asia*. New Delhi: Sage, 269–87

Ravindran, S. T. K. 1999. 'Female Autonomy in Tamil Nadu: Unravelling the Complexities'. *Economic and Political Weekly*, 34: 16–17, 34–44

Raz, J. 1986. *The Morality Of Freedom*. Oxford: Oxford University Press

Rothschild, S. C. 1982. 'Female Power, Autonomy and Demographic change in the Third World', in R. Anker et al., ed. *Women's Roles and Population Trends in the Third World*. London: Croom Helm

Sathar, Z., Crooke, N., Callum, C. and Kazi, S. 1988. 'Women's Status and Fertility Change in Pakistan'. *Population and Development Review*, 14: 3, 415–32

Sharma, M. and Vanjani, U. 1993. 'Engendering reproduction: The political economy of Reproductive Activities in a Rajasthan Village', in A. W. Clarke, ed. *Gender and Political Economy, explorations of south Asian systems*. Oxford: Oxford University Press

Stoler, A. 1977. 'Class Structure and Female Autonomy in Rural Java'. *Signs, Journal of women in Culture and Society*, Special Issue, 3: 1, 74–89

Street, B. V. 1984. *Literacy in Theory and Practice*. Cambridge: Cambridge University Press

Taylor, C. 1979. 'What's wrong with negative liberty', in A. Ryan, ed. *The Idea of Freedom*. Oxford: Oxford University Press

Unnithan-Kumar, M. 2001. 'Emotion, Agency and Access To Healthcare: Women's Experiences of Reproduction in Jaipur', in S. Tremayne, ed. *Managing Reproductive Life: Cross cultural themes in sexuality and fertility*. Oxford: Berghahn Books

Vlassoff, C. 1996. 'Against the Odds: The Changing Impact of schooling on female Autonomy and Fertility in an Indian Village', in A. M. Basu and R. Jeffrey, eds. *Girls Schooling, Women's Autonomy and Fertility Change in South Asia*. New Delhi: Sage, 218–34

Visaria, L. 1996. 'Regional Variations in Female Auotonomy and Fertility and Contraception in India', in A. M. Basu and R. Jeffrey, eds. *Girls Schooling, Women's Autonomy And Fertility Change in South Asia*. New Delhi: Sage, 235–68

Watkins, S. C. 1993. 'If All We Knew About Women was What We read in Demography, What Would We Know?'. *Demography*, 30: 4, 551–77

Watson, G. 1975 'Free Agency'. *Journal of Philosophy*, 72: 8, 205–20

Weller, R. H. 1984. 'The Gainful Employment of Females and Fertility: With Specific Reference to Rural Areas of Developing Countries', in W. A. Schutjer and C. S. Stokes, eds. *Rural Development and Human Fertility*. London and New York: Collier Macmillan, 151–71

Wolfe, B. L. 1980. 'Child bearing and /or Labour Force Participation: the Education Connection', in J. L. Simon and J. Da Vanzo, eds. *Research in Political economics*. Vol. 2. Greenwich Connecticut: Jai Press

NOTES ON CONTRIBUTORS

Monica M. E. Bonaccorso completed a Ph.D. in Social Anthropology at Cambridge under the supervision of Marilyn Strathern on 'The Traffic in Kinship: Assisted Conception for Heterosexual and Lesbian and Gay Couples in Italy'. Her work is on the use of kinship idioms in the context of assisted conception in Italy and how these relate to idioms from the North European/American context. It offers an unusual perspective through a comparison between heterosexual couples suffering from impaired infertility, lesbian and gay couples, and providers of treatment. The doctoral work in Cambridge builds on previous training in Philosophy at the University of Milan, where she earned a Laurea in 1993 with a dissertation on the bioethical questions surrounding lesbian and gay parenthood. This was subsequently published in 1994 by Editori Riuniti. Monica has published widely on culture, society and science working as a professional journalist in Italy. She has recently been awarded a three year Fellowship by the Wellcome Trust for a new project titled 'Cultures of New Genetics: An Exploration of Shared Idioms in the Media, Science Exhibitions and the "Public" in the UK' for which she will be based at Cambridge.

Henrike Donner is a Research Fellow at the Department of Anthropology, London School of Economics and Political Science. Her research interests include gender relations and kinship in South Asia, the impact of economic liberalisation and globalisation and contemporary politics in urban India.

Saraswathy Ganapathy is a paediatrician/ neonatologist with a Master's degree in Public Health. She is associated with the Belaku Trust, a nongovernmental organisation in Bangalore, India which carries out research and programmatic interventions in the fields of health and development. Her special interests are social and cultural determinants of women and children's health and nutrition.

Asha George is a public health researcher trained at Harvard and is currently based at the Institute of Development Studies in Sussex, where she is working on a project on women's activism and health in southern

India in collaboration with Professor Gita Sen of the Indian Institute of Management in Bangalore. Along with Professor Sen, she has recently coedited a volume on gender and public health.

Asha Kilaru is based in Bangalore, India and is involved in community based research in women and children's health with the Belaku Trust. She is interested in linking research to direct community work and understanding health outcomes within the context of women's empowerment.

Sumi Madhok has recently completed her doctorate at the department of political studies, School of Oriental and African Studies, University of London. Her dissertation is titled: 'Autonomy, Subordination and the "Social Woman": examining rights narratives of rural Rajasthani women'. Her other publications include: 'Autonomy, political rights and the "social woman": Towards a politics of inclusion' in Basu Subho and Crispin Bates (2002) (eds) *Rethinking Indian Political Institutions*, Anthem Press, London and 'Alternative Imaginings, Alternative Lives?' in, Sasha Roseneil and Linda Hogan (eds) *Feminism, Ethics and Agency* (forthcoming).

Shanti Mahendra has been a researcher with the Belaku Trust for some years and has co-authored several papers on maternal, child and adolescent health. Her interests are gender and empowerment/ autonomy issues in developing countries. She is currently pursuing an M.Phil. in Development Studies at the Institute of Development Studies at the University of Sussex.

Zoe Matthews is a Senior Lecturer in Demography at the Department of Social Statistics, University of Southampton. She is a founder member of the 'Opportunities and Choices' Research Group, a knowledge programme in reproductive health funded by the Department for International Development. Her research interests are primarily focused upon maternal health in Asia, but she has also worked on research projects on child health and contraceptive use in Africa and the Arab world.

Tulsi Patel is Professor of Sociology and head of department in the Department of Sociology, Delhi School of Economics, University of Delhi, in India. Her research interests are in the area of gender and population, population policies, fertility studies, anthropological demography, childbirth cultures and the related medical and therapeutic knowledge systems, medical sociology and anthropology, family and aging studies. She is the author of *Fertility Behaviour: Population and Society in a Rajasthan Village*, published by Oxford University Press. She had compiled and edited, *The Family through Abstract and Lived Categories*, to be published by Sage Publications as part of the golden jubilee series of the Indian Sociological Society, of which she was the treasurer for a few years. She has published several other papers on gender and fertility, and childbirth in journals and edited books in India and elsewhere.

Jayashree Ramakrishna is additional Professor and Head of Department of Health Education at the National Institute of Mental Health and Neuro-sciences (NIMHANS) in Bangalore, India. She is a medical anthropologist with a training in public heath education. She is the project coordinator for the Ford Foundation funded NIMHANS small grants programme for research on sexuality and sexual health behaviour. Her interests include reproductive health, sexuality and research methodology. She has been associated with the Belaku Trust since its inception.

Alison Shaw is a lecturer in Social Anthropology in the Department of Human Sciences at Brunel University, England. Her publications include, *Kinship and Continuity: Pakistani Families in Britain*, Harwood Academic Publishers, (2000). She is currently engaged in a project funded by the Wellcome Trust to investigate the impact of genetic risk information on families of Pakistani origin referred for genetic counselling.

Bob Simpson's doctoral research was carried out in Sri Lanka where he studied the transmission of skills and knowledge associated with rituals of exorcism and healing. His research interests thereafter shifted to the U.K. where he was involved for almost ten years with a series of projects investigating the process of divorce and separation with particular reference to dispute resolution and management. He has authored and co-authored numerous articles and reports on issues such as divorce conciliation, housing and divorce and the experience of fatherhood following marital breakdown. In 1998 he published a monograph which considered divorce and separation from an anthropological perspective (*Changing Families: An Ethnographic Approach to Divorce and Separation*, Oxford: Berg). More recently his interests have turned to the new reproductive and genetic technologies in the UK and particularly the way that these technologies are changing understandings of kinship, identity and personhood. In summer of 2000 he embarked on a new project which explores the way that the new technologies are currently being received in Sri Lanka. He has recently been awarded a Wellcome Fellowship to pursue this research in greater depth.

William Stones graduated in medicine in 1979 and worked on a community health project in India with a non-governmental organisation before pursuing specialist training in obstetrics and gynaecology in Britain. He now holds the post of Senior Lecturer at the University of Southampton and Consultant at the Southampton University Hospitals NHS Trust. As well as running a regional multidisciplinary referral service for women suffering from chronic pelvic pain and pursuing research in this area he maintains a research focus on reproductive health in the developing world through the 'Opportunities and Choices' programme based at the University of Southampton.

Soraya Tremayne is the Co-ordinating Director of the Fertility and Reproduction Studies Group, University of Oxford and a Research Associate at the Institute of Social and Cultural Anthropology, University of Oxford. She was Vice-President of the Royal Anthropological Institute. Her current research interests are in the sexual and reproductive health and behaviour of young people in Iran.

Maya Unnithan-Kumar is Senior Lecturer in Social Anthropology at the University of Sussex. She completed her doctorate from Cambridge University in 1991. Her doctoral work focused on kinship, gender and family relations in a poor 'tribal' community in North-West India. This was published as a monograph, *Identity, Gender and Poverty: New Perspectives on Caste and Tribe in Rajasthan* (1997). Her current interest is in the field of medical anthropology, especially in the anthropology of reproduction and health. She has published several articles on the issues of emotion, knowledge and agency in relation to childbirth processes, reproductive entitlements and health. She is currently working on two projects in the area of childbearing in Rajasthan: one related to migration, and the other with regard to ayurvedic medicine.

INDEX

A
Abbasi-Shavazi, M. J. 186
Abortion 5, 13, 14, 44, 122, 185, 190, 122, 185, 205, 206
   clinical assumptions about Muslim views on 35
   ethical dilemmas of 37–38
   Islamic positions on 36
   local British Pakistani views on 35–38
   preference for early 37
   services 74
   silence concerning 36
   in Sri Lanka 47, 48, 53, 110
   voluntary miscarriage 70
adolescents 116, 182, 183, 188, 189, 190, 192, 193, 194, 196
adoption 11, 44, 49
   in Sri Lanka 51–52
affines, affinal 12, 59, 66, 69, 70, 72, 75, 120, 124, 125, 128
agency viii, 1, 6, 7, 9, 15, 17, 64, 115, 128, 130, 213, 223, 224
   doctor's 5
   medical 103
   reproductive 69, 212
   spiritual 9, 64
   women's 15, 227, 229, 236
ambivalence 4, 9, 10, 68, 69, 103
   maternal 65
amniocentesis. *See* sex selection
anaesthesia/anaesthetic 89, 108
analgesia 108, 109
antenatal care 122, 124, 169, 215, 216
artificial insemination (AI) 49, 52, 84, 87, 95
   by donor (AID) 95, 49
autonomy 7, 14, 15, 17
   reproductive 6, 108, 124, 223–25, 227–37, 241
   doctor's 144
   female/women's 15, 207, 208
   scientific 153
Ayurveda 7

B
Baeta 124, 132
bananthana 13, 161, 162, 164, 167–69, 171, 174–77
banjhpan 64
Basu, A. M. 206, 207, 214
Belizán 121, 126
Bhatia, J. 162, 163, 165
bioethical 8
biogynaecological/ist 60, 63
biomedical 1, 4, 7, 8, 10, 61, 62, 63, 64, 65, 68, 69, 70, 71, 72, 73, 75, 103, 143, 153, 162, 177
birth 40, 60, 68, 69, 70, 118, 119
   after 12, 13, 30, 32
   approaches to 76

breech 68
by Caesarean section 38
canal 167, 169, 204, 205, 206
engineering 10
home vs hospital 9, 113, 116, 121, 123, 124, 125, 126, 127
kinship 94
local healers 74
medicalisation of 4, 123, 129
'normal' 114, 122, 126, 130
traditional 124
pain 109, 118
pollution 126
rate 11, 116, 184, 186, 187
surrogate 10
body/bodies 4, 5, 6, 8, 38, 48, 61, 65, 88, 89–92, 104, 105, 122, 123, 124, 128, 130, 161, 167, 169, 174, 176, 177, 205, 206
Bonacorso, M. 8, 11, 17
bonding 12, 65, 66
maternal 65, 66, 75
Bourdieu, P. 90, 96, 130, 203, 204, 210, 212
Boyden, J. 196
breast feeding 8, 176, 185
breech. *See* birth/birthing
buddhist clergy (*sangha*) 45, 47

C

caesarian section 2, 8, 11, 38, 108, 110, 121
Calcutta 4, 8, 11, 13, 119
introduction of hospital births 114
urban middle class 113, 116
caste 5, 6, 51, 63, 68, 71, 73, 75, 115, 125, 165, 206, 209, 210, 213, 215
and adoptions 51
and discrimination 73
and doctor-patient relationships 66
low 61, 64, 65, 124
and medicalisation 72
upper 121, 123
Chatterjee, P. 116, 120, 121, 123
Chennai 115, 126
Childbirth. *See* birth/birthing
chromosomal abnormalities 25, 34
chronic pelvic pain 109, 111

class 5, 8, 13, 64, 66, 67, 69, 72, 73, 114, 120, 165, 210
middle 4, 8, 11, 13, 67, 72, 73, 113, 114, 115, 116, 120, 121, 122, 124, 126, 128, 129, 130, 219
upper 205, 115
working 73, 115, 124, 126, 128
clinical genetics 25–26
facilities for, in Pakistan 30
reasons for referral to 27
commoditisation 195
confinement
food 125, 169
postpartum 118, 124, 125
consanguinity 26, 30, 32, 35, 37, 39, 40
consumption 62, 68, 69, 114, 128, 129, 147
contraception 139, 141, 145, 147, 149, 165, 170, 171, 183, 185, 186, 205, 206, 211, 215
devices 9, 15, 16, 47, 61, 71, 72, 91, 113
Convention of the Rights of the Child (CRC) 194
Correa, S. 16
counselling 4, 7, 14, 26, 33, 40, 145, 183, 189, 190, 191
Csordas, T. 6, 10

D

Dai 124, 125, 169
Davis-Floyd, R. and Sargent, C. 2, 7, 62
deafness 28, 31
demography 2, 14, 16, 21, 73, 203, 205, 206, 209, 213, 217, 218, 227, 228, 229, 231
development 2, 14, 16, 203
literature 6, 12, 14
organisation 3
desire (emotion) 9, 16, 145, 214, 219, 227
Dickey, Sarah 115
dilatation and curretage 59, 63, 65, 70–72, 74, 75
divining 63
division of work 128, 129
doctors
attitudes of 68, 72

private 73
role in hospital births 70
status 68
dominant inheritance. *See* genetic disorders
Donner, H. 6, 8, 11, 12, 13, 68, 73, 113, 115, 128
dysmenorrhoea 110, 111
dyspareunia 107

E

early marriage/s 165, 190, 191, 192
Edwards, J. 1, 2, 6, 9, 10, 17, 114
Electroencephalogram 33
endometriosis 107, 109, 110, 111
epidural anaesthesia 108
ethics/ethical 2, 4, 17, 52, 54, 138, 143, 150–53
  bioethics 8
  medical 16, 44, 51
ethnodemography 17
etiology 7, 175

F

family planning 13, 16, 71, 73, 145, 176, 182–86, 188, 189, 190, 191, 197, 206
  in Sri Lanka 45, 47
fertility 1–4, 12, 14–16, 71, 72, 73–75, 89, 91, 103, 111, 123, 145, 165, 191, 192, 197, 198, 206, 207, 210, 215, 217, 227, 228, 229, 230, 231, 233, 234, 235, 237
  male 11, 14, 47, 48
  private clinic 108
  rates 182, 186, 189, 205
  in Sri Lanka 46
  state control 3, 16
  studies 203
  and women's autonomy 14–15
foetal anomaly screening
  clinical views of 35
  to obtain information 35, 36
foetal scanning 63, 64, 70
food
  before birth 123–24, 125, 167
  after birth 7, 169, 174, 177
Foucault, M. 3

G

gender 1, 5, 43, 60, 67, 69, 72, 75, 103, 105, 114, 130, 141, 193, 195
genetic
  counselling 4, 7, 14, 26, 33, 40, 145, 206, 207, 208, 210, 211
  disease 39
  disorders 14, 34
    among British Pakistanis 26
    British Pakistanis understandings of 27, 29, 30–31
    dominantly inherited, clinical definition of 25
    generational differences in understandings of 31–32
    recessively inherited, clinical definition of 25–26
  origins 52
  risk 2, 39, 40
    clients' understandings of 27, 32–34
    clinical definitions of. *See* genetic disorders
    and social class 26
  technology 8, 44, 54
  testing 68
George, A. 9, 16, 137
Georges, E. 26, 60, 61, 66, 67
Ginsburg, F. D. 1, 2, 63, 91, 114, 130
globalisation 113, 193
Greenhalgh, S. 2, 15, 73, 215, 228, 236

H

Haeri, S. 195
health
  care 30, 111, 130, 145, 163, 170, 171
  insurance 129
  private 129
  policy 13, 14, 15, 16, 182, 183, 192, 196
  reproductive 1, 2, 3, 14, 16, 49, 70, 113, 127, 137, 145, 182, 183, 186, 188, 189, 190, 192, 194, 196, 197
  services 10, 71, 145, 164
Hindu, caste 45, 47, 61, 64, 66, 71, 72, 75, 214
home birth 113, 123
Hoodfar, H. 184, 185

hospital
  as modern institution 27
  as place of birth 8, 9, 11, 12
households 51, 53, 68, 72, 115, 122, 128, 129, 165, 210, 211, 215

I
ICPD 183, 188, 216
ideology 70, 116, 193, 198, 205
  gender 67, 210
  Islamic 184
  patriarchal 66, 69, 71
immunisation 13, 170, 171, 176, 177
individualism 10, 15, 223, 224
infant mortality 30, 45, 60, 175, 186
infertility, sterility 12, 59, 62, 64, 72, 74, 75
  male 48, 49, 50, 52, 53
  treatment 83, 84, 85, 88
Inhorn, M. 2, 16, 59, 74, 114, 129, 211
intrapartum care 108, 162
intra-uterine device 8, 18, 71, 74, 122, 138, 185
in vitro fertilisation 6, 11, 43, 84
  in India 44
  in Sri Lanka 49
IPPF 188
Islam/Islamic 7, 14, 188, 190, 194, 195, 196, 197. *See also* religion, *pirs*
Islamic Republic of Iran 13, 181, 182, 184, 186, 188, 194, 197

J
Jain 139, 141, 142, 143, 144, 146
Jaipur 60, 61, 62, 64, 70, 71, 128, 226
Jasanoff, S. 149, 150, 151, 152, 153, 156
Jeffery, P. 127, 128, 231
Jeffery, R. 214
Jordan, B. 7, 60, 62, 127

K
Kandyan Law 51
Kaufert, P. 2, 4, 9
Kessel 137, 139, 141, 142, 143, 147
Khomeini, Ayatollah 184
Kielmann, K. 2, 15, 59, 74
Kilaru 6, 7, 12, 13, 17, 161

Kini 140, 141, 142, 146
kinship 2, 10–12, 43, 44, 49, 51, 52, 53, 69, 94, 95, 97, 120, 207, 213, 217
kinswomen 62
knowledge 62
  authoritative 2, 7, 9, 15, 60, 62, 75, 127, 137, 140, 142, 144, 147, 148
  embodied 68
  medical 7, 43, 116, 128, 130

L
labour 60, 68, 108, 109, 118, 125, 127, 165, 194, 195, 210, 214, 238
laparoscopy 110
Leach, E. 51, 53, 55
learning difficulties 28, 29, 31
life-expectancy
  in Sri Lanka 45
life-support technologies 28, 38–9
Lock, M. 2, 4, 5, 9, 44
Lyon, A. 128, 218, 231

M
Madhok, S. 6, 15, 205, 206, 214, 223
Malthus 205
Martin, E. 1, 3, 4, 6, 48, 60, 66, 72, 123
maternal
  health 13, 47, 175, 178
  histories 115, 123, 129, 130
  morbidity 162, 163, 177
  mortality 16, 121, 141, 142, 147, 162, 178, 192
medicalisation 2–4, 5, 72, 88, 89, 108, 115, 129
medical language 8, 93, 94
medical termination of pregnancy. *See* miscarriage, voluntary
men 1, 3, 6, 8, 18, 46, 48, 53, 59, 60, 62, 63, 72, 75, 95, 107, 113, 125, 186, 195, 204, 205, 208, 213, 215, 216, 218, 227, 228, 233, 238, 240
Mernisi, F. 195
metabolic disorders 29, 30
midwives 8, 9, 12, 60, 61, 108
  auxiliary nurse 63
  categories of 62, 63, 64, 68, 69, 75
  local 61
Mir Hosseini, Z. 195

miscarriage 60, 62, 64, 67
  involuntary 63, 70, 71
  voluntary (abortion) 70
Mitchell, J. 60, 61, 67
mobility 55, 114, 123, 128, 162, 163, 168, 175–77, 208, 209, 215, 233
modernity 1, 13, 68, 123, 184, 192, 198
modern medicine 61, 64, 85, 91
Moghadam, F. 195
Morgan, S. P. 228, 232
multigravidas 108
Mumford, S. 139, 142, 147
Muslim, Sunni 61–63, 68, 71, 72, 75

N
neonatal 27
neurophysiological 104
nociception 106

O
obesity 29, 31
obstetrician 9, 27, 103, 110, 119
overpopulation 184

P
paediatrician 27, 33
Pai, M. 115, 121, 126
pain 9, 65, 69, 73, 173, 174
  childbirth 107–110, 118, 119, 126
  chronic 104, 105, 107
  inflammatory 104
  International Association for the Study of 104
  management of 89, 126, 127
  menstrual 105–107, 109
  neuropathic 104
  pelvic 107, 109, 111
  perception 103
  recall 108, 109
  relief 72, 108
  visceral 104
Panda, M. 2, 15, 74
Papanek, H. 128
Patel, T. 6, 15, 73, 122, 125, 203, 208, 209, 210, 212, 214, 215
patient
  birthing woman as 88
patrilocality 123
perinatal 27

personhood, foetus 8, 49, 54, 59, 60, 230, 247
Petchesky, R. 1, 4, 5, 8, 16, 17, 60, 66
Pigg, S. 12, 13, 62, 73
pirs
  as healers 28, 31
  as moral advisors 38
pollution 118, 126
  dai as removers of 124
  and midwives 63
  past management of 125
  shame 125
polyandry in Sri Lanka 11, 51, 53
population planning 3, 15, 16
postpartum period 1, 7, 11–13, 120, 124, 127, 130, 161–65, 167, 169–78
postnatal 7, 62, 125, 163, 170, 171
pregnancy 1, 7, 27, 29, 34, 35, 36, 37, 39, 47, 66, 68, 69, 70, 71, 87, 108, 121–23, 124, 161, 162, 163, 167, 171, 175, 183
prenatal 35, 65, 66, 68, 162
prenatal diagnosis. *See* foetal anomaly screening
primigravidas 108, 163
private/privatisation 17, 71, 72, 74, 75, 84, 85, 87, 88, 97
  effects on healthcare in India 113–15, 119–22, 124, 131, 139
  and 'elective' Caesareans 113
  in the wake of liberalisation 113
public health 2, 13, 14, 15, 65, 71, 73, 138, 139, 140, 143, 146, 147, 148, 153

Q
quinacrine sterilisation 4, 9, 16, 138, 139, 140, 142, 144, 146, 148
Qur'an 31

R
race 5, 14, 72, 114
racism 14, 26, 29, 30–31
Ragoné 1, 2, 4, 5, 10, 17, 84, 89
Rajasthan 59, 60, 61, 62, 65, 66, 68, 74, 75, 203, 211, 226, 236
Rao, M. 17, 145

Rapp, Rayna 2, 5, 26, 44, 60, 63, 66, 69, 72, 73, 114, 130
Ravindran, T. K. 2, 15, 74
recessive inheritance. *See* genetic disorders
religion 7, 14, 196
   and birth control 27, 36
   as a framework for ethical decision-making 38–39
   and prenatal diagnosis 27, 34, 35
   and termination of pregnancy. *See* abortion
   versus scientific knowledge 36–37
reproductive, desire 62, 69
Riesman, Catherine K.
rights 2, 5, 120, 127, 137, 203, 204
   children's 194
   men's 60
   reproductive 16
   women's 12, 108, 145, 204, 208
Rozario, S. 62, 73, 124, 127

S

safai 63, 69, 70, 71
Sangari, Kumkum 128
sangha 45, 47
self/selves 5, 69, 228, 231, 232, 235, 236
Selfe, S. A. 106, 110
semen 48, 52, 53
sex selection/amniocentesis 8, 35, 36, 67, 70, 122, 123, 146
shadh 124
shakti 126
shame
   of pregnancy 123
   related to birth pollution 126
Shapiro, T. M. 205, 219
Sharia 182, 183, 190, 194
Shia 182
Shweder, R. 10
Simpson, B. 8, 11, 14, 43, 103, 120
Sinhala Buddhist 45, 46, 47
Sinhala Urumaya 46
social
   change 74
   connectedness 67
   intimacy (emotion) 62, 65, 71
   obligation 5
   relationships 59

reproduction 49, 52, 60, 61, 63, 73
responsibilies 68
science 151
State 3
status 67
values 3, 60, 70
violence 48
spacing, of children/birth 61, 75, 190
spiritual healers 7, 60–64, 68
Strathern, M. 1, 2, 4, 5, 10, 17, 84, 89
status 15, 60, 67, 68, 74
   social 15, 204, 205, 206, 207, 210, 212, 214, 217
   of men 208
   of women 122, 203, 206, 211, 213, 217, 218
sterilisation 4, 9, 16, 138, 144, 146, 147, 148, 165, 169, 170, 226
Stones, W. 9, 31, 103
Sunni Muslim 61, 62, 63, 75

T

Taylor, J. 5, 60, 61, 65, 66, 129
technologies 1, 3, 4, 10, 13
   genetic 8, 44, 54
   IVF 11
   medical 1, 17, 25, 39, 59
   reproductive 1, 2, 4, 5, 8, 9, 10, 11, 12, 13, 14, 27, 43, 49, 50, 51, 53, 62, 64, 66, 71, 74, 75, 76, 91, 114, 123, 127, 128, 129, 130
termination of pregnancy. *See* abortion
Theravada Buddhist 46
Tremayne, S. 13, 17, 181
tubectomy 13, 59, 71, 72, 74, 170, 171, 176

U

ultrasound scan 35–36, 59, 60, 63, 65–68, 72, 74, 75, 78
UNFPA 191
Unnithan-Kumar, M. 6, 8, 9, 10, 12, 13, 17, 62, 73, 74, 120, 122, 127, 128, 212

UNICEF 191, 194
urban 8, 60, 114, 115, 120, 121, 124, 126, 127, 128, 130, 144, 162, 186
Uttar Pradesh 128

V
Van Hollen, C. 13, 73, 126
viagra 48

W
Work 15, 28, 129, 130, 168, 175, 185, 210, 211, 215
World Health Organisation (WHO) 138, 139, 140, 143, 144, 145, 161, 164, 175, 177, 178

Z
Zipper, J. 138, 139